Reflections on Science and Evolution

Andrés Moya

Reflections on Science and Evolution

Springer

Andrés Moya
Institute for Integrative Systems
Biology (I2Sysbio)
University of València
València, Spain

ISBN 978-3-032-25681-2 ISBN 978-3-032-25682-9 (eBook)
https://doi.org/10.1007/978-3-032-25682-9

Moya, A. (2020). Ciencia en pequeñas dosis: Reflexiones sobre ciencia y evolución. Cálamo.

Translation from the Spanish language edition: "Ciencia en pequeñas dosis. Reflexiones sobres ciencia y evolución" by Andrés Moya, © The Author 2020. Published by Ediciones Cálamo. All Rights Reserved.

This work was supported by Universitat de València (Sabattical Leave 2019), Fulbright Program (Sabbatical leave 2019) and Ministerio de Ciencia, Innovación y Universidades (Sabattical leave 2019).

This Springer imprint is published by the registered company Springer Nature Switzerland AG
The registered company address is: Gewerbestrasse 11, 6330 Cham, Switzerland

Foreword

Although its title suggests that it is a work of popular science, this book is much more than that. Interest in science has historically not been a characteristic of Spanish society. Our scientific culture is limited, and science is still not considered an integral part of our broader cultural identity. But, little by little, popularization is helping to change this situation. Science is increasingly present in society. Increasingly, initiatives are being launched to make science accessible to the public, enjoyable, understandable, and valued. And many scientists finally recognize that their reputation does not have to be diminished (quite the contrary) by devoting part of their time to popularization. Professors and researchers are beginning to appear regularly in the media, write books, publish blogs, or create monologues aimed at capturing our interest, entertaining us, sharing the history of science and its anecdotes, and, in short, helping citizens overcome their fear of science and see it as an essential part of culture.

This book fully achieves all those objectives of popularization but goes further. In one of the chapters, Professor Andrés Moya sets out his vision of what he calls the "triumvirate of science." In it, he stresses the importance of popularization, but recognizes that the entertainment component—designed to attract attention and foster enjoyment of scientific advances—is not enough. Outreach is a complementary activity that must be built on two pillars: education and reflection. Popularization should complement a solid scientific education for young people—an education that fosters a critical spirit, rejects dogmatism, stimulates curiosity, and teaches us to live with uncertainty and doubt.

However, perhaps the most significant difference between this work and typical popular science books is that *Reflections on Science and Evolution*, apart from entertaining and educating, invites us to think. Throughout the book, the author reflects (which, in his own words, means "to think carefully and attentively about something") on many subjects. And that is the effect he also generates in the reader. It is an extraordinarily entertaining and easy-to-read book. The author's ability to express his ideas in an organized, clear, and effective way is evident. But behind this accessible language lie concepts of great strength and depth. I recognize that, on many occasions, reading this book has generated the same feelings as reading the texts of Ortega y Gasset, a philosopher Professor Moya admires and considers a master. As with some of Ortega's works, the reflections in this book were previously published in the press—specifically, in columns over two years in the Mexican weekly *El Ángel Metropolitano*. This fact highlights Professor Moya's commitment to bringing science closer to non-specialized readers. As with Ortega, the author strives to write with clarity ("the courtesy of the philosopher," as the master used to say). But perhaps the most significant similarity lies in the fact that, in both cases, these are simple texts that nevertheless contain profound ideas—ideas that make the reader pause from time to time to reflect on what they have just read. Those moments of pause, reflection, and connection with the author are the book's greatest treasures. They are moments of deep enjoyment.

Therefore, more than merely an informative work, this is a collection of science essays. And, as with its title, the book offers much more than its subtitle promises. It is not only about reflections on science and evolution, nor is it solely about evolutionary biology—Andrés Moya's specialty. A significant portion of the chapters is devoted to reflecting on the essence of science, including its definition, objectives, and particular characteristics. The author argues that science is the best tool for understanding the world around us, for gaining insight into the unknown, and for gradually approaching the truth—a truth that is often unattainable, which makes scientific knowledge both limited and infinite. However, his staunch defense of science and his strategy of trying to analyze the world "from science" do not lead him to scientism. He consistently defends the need to resort to forms of knowledge beyond science, which help us understand the world from other perspectives.

The author's erudition and interest in other fields of knowledge are evident throughout the volume—and perhaps one of its greatest attractions. The chapters analyzing the relationship between science and philosophy are truly outstanding. These reflections are difficult to imagine from an author who is not both a scientist and a philosopher, as Andrés is—a double vocation that emerged when he was still a child, spending his days doing experiments and trying to understand the world. This vocation later led him to pursue two degrees, in biology and philosophy, because he saw that the first could not answer all his questions. Throughout the book, he defends the study of philosophy not as an alternative to other types of knowledge but as a foundation before any specialization. "Let us learn philosophy!" he proclaims, because studying it is the best remedy against indoctrination. The distance between dogma and skepticism is a common element in science and philosophy—two disciplines based on reason that have converged in recent decades, yet maintain their distinctive characteristics.

His integrative vision is not limited to science and philosophy. He finds little sense in the early separation of educational paths when choosing between the scientific, the humanistic, and the artistic. In his texts, he sets a good example. From a scientific perspective, he constantly reflects on politics, history, language, psychology, art, music, mathematics, and a long list of topics, too extensive to enumerate in a short foreword like this. One of the most inspiring aspects of this essay is his devotion to science and politics, particularly his defense of the right of individuals to participate in scientific endeavors. On the one hand, benefiting from technological advances, but above all, the fundamental right to scientific education. "Nothing is more liberating than education and culture," he assures, and in the magnificent chapter "Science for a Participatory Democracy," he maintains that "scientific instruction is essential to shield us from demagogy, to make us free citizens and critically engaged participants in decision-making within a participatory democracy."

I have been fortunate enough to know Andrés Moya for years. As science, medicine, and humanism are the three pillars on which all the initiatives of the Fundación Lilly are based, it was inevitable that our paths would cross at some point. And they have done so on several occasions. For example, he is the only scientist to have received the two most prestigious awards of the Foundation that I have the honor of directing. In 2013, he won the Lilly Foundation Award for Biomedical Research in its preclinical category, recognizing his contribution to a better understanding of the microbiome. In 2019, he was named Master of the Andrés Laguna Lecture. This award recognizes the work of Spanish and Latin American scientists who have distinguished themselves through their scientific and humanistic contributions. His Master's Lecture, delivered at the Paraninfo of the University of Alcalá, was entitled "I'm Me with My Microbes," a phrase that, besides being a clear allusion to his mentor, reflects the author's characteristic taste for the crossbreeding of knowledge. It is an absolute honor for the Fundación Lilly to have contributed to the publication of a book like this, which is likely one of the finest science essays written in our language.

In Spain, we usually complain about the lack of cultural references. Professor Andrés Moya is a clear example of the fact that cultural references do exist. The problem is that our society does not recognize them. The book's content was first published in a Mexican weekly, but it could also serve as a core reference for many university programs in science and literature. I sincerely believe that it has nothing to envy compared to such famous books as *What Is This Thing Called Science?* by Alan Chalmers, which has been used for generations in the education of university students worldwide. It is incomprehensible that an intellectual of his stature does not have a regular space dedicated to science in one of the country's leading media outlets. Our Mexican friends had the good fortune to read him weekly for two years. Hopefully, this book will serve as a stimulus, ensuring the initiative of El Ángel Metropolitano continues in the rest of the world, and we can all continue to enjoy the wealth of knowledge that Professor Andrés Moya still has to share.

This foreword does not aim to list all the author's achievements. Suffice it to say that, among many other merits and occupations, he is a Professor of Genetics at the University of València, directs the Institutional Chair FISABIO–University of Valencia, and is the author of more than six hundred publications on genetics, evolution, and evolutionary thinking; he has won numerous awards, including the National Genetics Prize in 2012, the Mexico Science Prize, and the aforementioned Lilly Foundation Research Award. He has been a visiting professor at the University of Irvine in California, and more recently, he enjoyed a sabbatical stay in the Department of Evolutionary Biology at Harvard University. During this period, he wrote a significant portion of the book. He shares with us his experiences in fragments of the book. After reading these texts, we can imagine the professor devouring all kinds of books during his subway rides, thinking deeply, with that melancholy he says scientists feel “when their natural inability to understand or make sense of their existence becomes omnipresent.”

Lilly Foundation
Madrid, Spain

José Antonio Sacristán,

Competing Interests The author has no competing interests to declare that are relevant to the content of this manuscript.

Introduction

In January 2018, I was invited to write a column for the Mexican weekly *El Ángel Metropolitano*. It was an activity I committed to consistently for two years, during which I reflected from many angles on science as a whole and on biological and cultural evolution in particular. These columns form the core material of this book, but they have been extensively revised, updated, and thematically organized into six sections.

Why "Reflections"? To "reflect" means to think carefully and attentively about something. And that is precisely what I do here—on science (Chaps. 1–4) and natural evolution (Chaps. 5 and 6). The reflections within each section are presented in the chronological order in which they were written, and each can be read independently.

In Chap. 1, I explore what we call science and its particularities in relation to other forms of knowledge about the world. I argue that science is always an unfinished and evolving project, which is why it is infinite in scope. Viewed from many angles, its approximate nature, distinct from other forms of knowledge, emerges as its defining characteristic.

Chapter 2 brings together a series of reflections that offer an internal perspective on science, treating it as a social institution. Although it has its own structure and function, science shares the strengths and weaknesses common to other human institutions.

In Chap. 3, I examine the long-standing debate about the relationship between science and other domains of knowledge—namely, philosophy, theology, and the arts. Having accepted that science must engage with these different forms of knowledge in addressing the great questions of existence and meaning, I argue for a relationship of equality and dialogue among them. I also critique the scientistic dogmatism that arises when science assumes itself to be the only valid mode of thought.

Chapter 4 forms a central part of the book's reflections. Here, I argue that science and science education should be considered a fundamental right of citizenship. I also reflect on the crucial role of science, in collaboration with national and supranational political leadership, in advancing humanity and safeguarding the planet. I emphasize the importance of defending and recognizing Ibero-American science and other cultural spheres, in contrast to the dominance of Anglo-Saxon science. Although science is often described as universal and beyond geography, I argue that its universality should not be geographically neutralized.

Chapters 5 and 6 offer a broad range of reflections. Chapter 5 focuses on biological evolution as a historical process that can be revisited and from which human society has much to learn. Chapter 6, under the umbrella of the "artificial"—understood as the capacity to intervene in nature—presents a series of reflections on the future of our species, including the ethical and ontological consequences of modifying ourselves and creating entities that seek to emulate us.

I want to thank Prof. Amparo Latorre and Dr. Antonio Moya Latorre for the critical and systematic review of all my reflections. I also extend my gratitude to the Universitat de València, the Ministry of Universities, and the Fulbright Foundation for the grants that supported my sabbatical at the Department of Organismic and Evolutionary Biology at Harvard University (Cambridge, USA) during the first semester of 2019. I am especially grateful to Prof. Gonzalo Giribet of the department mentioned above for welcoming me into his group and for his generous hospitality, which provided an exceptional environment in which to develop a significant portion of this book.

Contents

Chapter 1
What Is Science?

Abstract In this chapter, I reflect on the nature and value of science as a way of thinking and understanding the world. These sixteen texts explore science's relationship with truth, its resistance to dogma, and its openness to uncertainty. I argue that science is not just a body of knowledge, but a cultural and intellectual pursuit that belongs to everyone. I examine how scientific thinking can shape democratic societies and why it must be communicated clearly and responsibly. Through personal reflections and examples, I advocate for a deeper public engagement with science—one that goes beyond entertainment and fosters curiosity and critical thought. My goal is to make science accessible without losing its complexity, and to encourage readers to think carefully and attentively. These texts are written in a direct and engaging style, inviting reflection on ideas that are both profound and urgent. These reflections are not intended as definitive claims, but as invitations to examine the strengths and limits of scientific knowledge with intellectual humility and openness.

Keywords Complexity · Democratic societies · Nature · Scientific thinking · Think carefully · Value

1.1 Science, Always More Science

In my book *Pensar desde la ciencia* (Moya 2010), I argue that "alongside other forms of discourse, science adds the capacity to see and transform the external world—like the one glimpsed from Plato's cave." We cannot, nor should we, ignore science when facing the challenge of continuing to exist. Other forms of knowledge and belief are also necessary. Still, science remains a truly vital instrument for ensuring our survival and that of the world we inhabit in all its dimensions. Why do I say this? In his course *Introduction to Philosophy*, Martin Heidegger (1996) comments that science seeks to know and explain the things of the world—the entities—to uncover them, but not to grasp their essence, a task more appropriate to philosophy or art. He also reaches two essential conclusions. First, science aims to dominate,

A. Moya, *Reflections on Science and Evolution*,
https://doi.org/10.1007/978-3-032-25682-9_1

manage, and direct the entities it investigates. Second, technology is an intrinsic part of epistemology; in other words, knowledge gives rise to techniques that allow us to control and master the things of the world.

The task of science differs from other forms of knowledge because it is never complete. It is always in a state of continuous progress and development. Therein lies both its greatness and its limitation. It is limited, yes, but also infinite. That is the conclusion reached by David Deutsch, a pioneer in quantum computing, who in his excellent book *The Beginning of Infinity* (2011), tells us that science has unlimited potential for the creation of knowledge—that it is a dynamic system that promotes both the growing explanation of entities and their control and mastery. While it is true that we must avoid, as Jacques Monod stated in *Chance and Necessity* (1971), "any confusion between the ideas suggested by science and science itself," it is equally valid that we must not avoid being educated in science or thinking from it. Because of its unique nature and profound relevance, science—and the thinking it fosters—constitutes, alongside other forms of knowledge, a key element in the framework of personal culture.

In this book, I intend to present scientific findings and reflect on their scope and implications. For example, how Darwin led us from speculation about biological evolution to solid evidence of its existence; how suspicion about a genetic basis for lethal violence in our species has turned into strong proof; how we are getting closer to predicting evolutionary processes, even though we know there is a constant tension between contingency and necessity in the evolution of biodiversity; and how the dream of manufacturing life is becoming a reality through synthetic biology. This defense of science should not be read as a denial of other forms of understanding, but as a recognition of its distinctive epistemic power and practical relevance.

1.2 The Origins

Three fundamental origins concern us and justify science's engagement with them: the origin of the universe, the origin of life, and the origin of humans. Although science addresses them separately, it aims to formulate a unified explanation, as if this were a natural aspiration. I am referring to an explanation in the form of a theory that can account for the three origins mentioned above. Do we have such a theory? Certainly not, but we are getting closer.

Regarding the origin of the universe, a set of fundamental elementary forces has been formulated whose action has led to the unfolding of everything that has subsequently appeared—including life on our planet, possibly on others, and humans. But are these fundamental forces sufficient to explain the emergence of complexities such as life and humanity? Some physicists say yes, arguing that their theories are based on boundary conditions that make the appearance of such complexities feasible. My impression, however, is that these theories remain imprecise and uncertain.

Thus, questions such as "where exactly can we anticipate life or intelligent life, or when" have no concrete answer. Like a funnel, the physical theory of the cosmos unfolds its explanation, marking certain margins. Still, it lacks the precision needed to predict specific events that have occurred at least once in the universe's evolution. We can affirm that its explanations are still weak. From this, we can reflect once again on both the humility and the grandeur of science.

Do the theories of today's cosmologists differ from the ancient cosmogonies formulated by various civilizations—such as the pre-Columbian ones—throughout history? Yes, of course: some are scientific theories, and others are not. However, the extraordinary precision achieved by some of those civilizations in understanding the dynamics of the stars and other astronomical phenomena is surprising. Let me venture to suggest to the reader that the formulators of those cosmogonies—regardless of their societal function—and today's cosmologists respond to the same intimate need: the search for answers and the attempt to make sense of existence. Is this really what motivates today's scientists? They would probably say no, that they are only trying to explain the universe based on cold, hard science—something that, according to them, differentiates them from the cosmologists of antiquity and, in any case, from ordinary people who, at least occasionally, ask existential questions.

As I see it, one cannot so easily separate the search for scientific explanations from fundamental questions. Like primitive cosmologists and people in general, the scientist's quest for explanation is fueled—however unconsciously—by the need to find meaning in existence. Just as we use science to promote social welfare, we also use it to seek fundamental answers to pressing questions. That is what two scientists with very different dispositions and beliefs—Einstein and Gödel—reflected on almost daily during their morning walks to the Institute for Advanced Study in Princeton.

Final thought: Science contributes, like art, philosophy, or theology, to providing answers—or even to concluding that there are no answers—for certain questions. But it has a particular dimension that distinguishes it from these other forms of understanding reality: its method, which allows it to reach explanations that are not final, but progressive and approximate. In general, the work of the artist, philosopher, or theologian must be creative or oriented toward the creation of a system based on thought or faith. These individuals, using various tools, approach the creation of artworks and systems. By contrast, the scientist's originality lies in the discovery and approximate explanation of what exists, including art, philosophy, theology, and science itself.

1.3 On the Limits of Science

I will not enter into the ethical debate about whether there are issues that science should not investigate. That is a very relevant question, linked to the type of society in which science develops, and I will have the opportunity to reflect on it later. Here, I would like to discuss whether there are absolute limits to what we can epistemologically come to know through science. It appears that, in specific fields, science itself has recognized that there are matters that cannot be known. That is the case, for example, with Heisenberg's uncertainty principle, Gödel's incompleteness theorems, or the unpredictability described by chaos theory. These could suggest that science has established absolute limits on what we can know or explain. However, if we take a brief tour through the history of science, we find such astounding discoveries that cast doubt on that conclusion. Such limits reflect the current state of scientific understanding, not an intrinsic incapacity of science to advance beyond them.

Let us take two examples. First, Mendeleev's proposal of the periodic table of chemical elements, which anticipated elements yet to be discovered. Second, research into atomic particles, which continues to advance toward the formulation of ever more elementary particles—perhaps reaching those that are truly fundamental, not composed of smaller constituents. These are counterexamples to earlier conceptions of supposed limits: namely, that there were no more chemical elements beyond those already known when the brilliant Russian chemist proposed his table, or when the first atomic model of matter's fundamental composition—based on electrons, protons, and neutrons—was introduced.

Let us now consider Gödel's theorems. Indeed, they establish that propositions may arise within an axiomatic system that cannot be proved true or false using the system's own axioms. However, it is possible to formulate a new system that incorporates additional axioms, allowing for the demonstration of those propositions' truth. Yet new propositions will inevitably emerge that remain unprovable, leading to a kind of infinite regress.

I see science as a mode of approximate knowledge—never finalized—of reality. Compared to other forms of knowledge, science has the particularity of never offering a definitive vision of anything. In recent decades, there have been frequent claims that science has reached the limits of what it can know and that no new theories will emerge in certain fields. Such proclamations have consistently collapsed. Science is constantly generating new knowledge, making progress on the vast ocean of ignorance. I like to think of it as a continuous movement that, strictly speaking, proceeds by successive approximations—some of which revolutionize previous theories. There is no contradiction between my affirmation of successive approximation and the emergence of profound changes in what is known. What is systematic is the doing of science. Science is never satisfied; it is always ready to open new paths to gain ground on the unknown. Hence, speaking of science is intrinsically equivalent to speaking of frontier knowledge. Insofar as everything that remains to be known is infinite, we can also say that science is infinite. Therefore, it can be considered a mode of knowledge without limits. The endless need to know makes science limitless.

1.4 Following the Search for New Principles

The philosopher Thomas Nagel has published an influential book that aims to be a devastating critique of the materialistic reductionism underlying neo-Darwinism. The book is entitled *Mind and Cosmos: Why the Materialist Neo-Darwinian*

Conception of Nature is Almost Certainly False (Nagel 2012). The explanatory limitations that Nagel believes he sees in biological science—particularly regarding two of its most complex questions, the origin of life and the origin of consciousness in the human species—are, in my opinion, somewhat superficial and rather conventional. Nagel relies heavily on proponents of intelligent design, whom he defends because they have the right, from legitimate religious positions, to question issues that, in their view, current biology does not adequately explain. In particular, regarding the origin of life, Nagel resorts to the argument of its extreme improbability.

On the subject of consciousness and the presence in our species of categories associated with it—such as judgments, values, or freedom—Nagel argues that this is the area where he most strongly perceives the impossibility of a scientific explanation, if it is reductionist and materialistic in nature. For Nagel, the physical laws that govern the framework of modern science lead not to an explanation, but to a non-explanation of consciousness, given that it is strictly immaterial while also capable of acting on the physical components of our body. Whenever these kinds of issues arise, I always wonder why there is so much reluctance to accept material bases for complex phenomena, even if we currently lack sufficient explanatory resources to account for them.

On the other hand, Nagel does not claim or intend to criticize neo-Darwinism to embrace theism. He is not satisfied with an explanation based on the unintelligible actions of a Supreme Being. On the contrary, he calls for the search for new answers, remaining within atheism—though not a militant one.

At a certain point in his argument, Nagel acknowledges that the only way to provide a scientific explanation for the origin of life and human consciousness would be through a more fundamental principle in the universe's evolution. Only from this perspective can we say that a materialistic (scientific) explanation of these two major issues is possible. However, the fact that we do not yet have such a theory, or evidence of a fundamental principle driving complexity—whether in the origin of life or consciousness—does not exclude the possibility that future science might discover it.

It is true that, at present, such a principle—at least in biological evolution—runs counter to standard neo-Darwinian thinking, where the unpredictability and contingent nature of biological phenomena make it impossible to conceive of a trend. But without invoking higher intelligence, demonstrating a physical law that promotes complexification in the universe's dynamics would protect us from unscientific explanations. In my opinion, it is dogmatic to criticize the materialist position of science because it is reductionist in nature. Reductionism and materialism are two very different concepts. The discovery of new laws specific to certain levels of matter organization can overcome reductionism.

Conversely, materialism advocates or assumes the existence of elements, factors, or forces of a natural—not supernatural—type. The fact that some of them elude us today does not exclude the possibility that we may discover them in the future. The materiality with which science currently operates does not exclude the emergence of new explanatory materials. Science seeks to explain natural processes through principles or laws that have been discovered over time. Hence, its infinite character—it is a work constantly under construction. So why not revise its most fundamental structures from time to time? Raising these questions does not imply abandoning scientific explanation, but rather taking seriously its capacity for conceptual renewal.

1.5 New Worlds

Without delving into its historical or political dimensions, "colonize" refers to the establishment of a group of people in a territory far from their homeland to populate it and exploit its resources. Contrary to what might be assumed, the term is not derived from Christopher Columbus (1451–1506), the discoverer of America. Its etymology is quite different, and this is central to my argument. "Colonize" derives from the Latin *colonia*, which refers to a territory settled by non-locals. That term comes from *colonus*, a farmer and inhabitant of a place, which in turn comes from *colere*, meaning to cultivate and inhabit a place.

The discovery of the New World marked the beginning of colonization processes in the modern and contemporary eras. But colonizations existed earlier—in prehistory and antiquity. In other words, we humans are colonizing beings. We have never ceased to be so, and to a large extent, our success on this planet—measured in terms of survival and population growth—is probably due to the enormous curiosity that precedes the act of "going somewhere else," regardless of additional, not always laudable, motives. This capacity undeniably characterizes us.

The planet has become too small for colonization. There is practically nowhere left to explore. But colonizing means settling, cultivating, and exploiting a place's resources. Some areas may remain inadequately colonized for various objective reasons. Still, I believe it is worth considering that, as a species, we have made our presence felt everywhere, used the resources, and are now on the path to exhausting them. It is to the species that I turn to remind us that our curiosity leads us toward new worlds—worlds that are no longer on this planet. Modern and contemporary history underwent radical changes with the arrival of Columbus, who came into contact with previously isolated worlds. But now we face a new reality: the new worlds lie beyond Earth.

Years ago, I was a talk show host on a science program for Spain's national broadcasting network. On one occasion, we discussed public and private initiatives to send humans to other places in the universe. It was surprising to see how many people we spoke to were willing to make a one-way trip, for example, to Mars. Many had already applied to participate in private initiatives for that purpose, and some had been selected. When asked about their reasons for such a unique adventure, their answers were broadly similar: they felt no strong attachment to their current world, and the challenge of venturing to a radically new place gave them hope.

As various space agencies are already doing, we can send devices or robots to other planets or satellites in our solar system. That is important for evaluating terrain characteristics. However, it is not comparable—in terms of novelty in scientific and technological research—to the challenges involved in sending human colonists. In this sense, space exploration should be included in international science funding programs. It is a common cause that unites our species beyond borders and nations. Without neglecting the need to fund research that enables a sustainable life on Earth, we must also support scientific research and technological development aimed at sending humans to other parts of the universe—starting, of course, with those that are closer and more feasible to colonize.

But, as I said, the challenges are so extraordinary that just thinking about them induces vertigo. Yet not pursuing them goes against our natural curiosity. Launching such a program could once again increase public interest in science and attract the best talent to it.

1.6 In Praise of Stephen Hawking

Stephen Hawking and Leonard Mlodinow state on the first page of their book *The Grand Design* (2010) that "… philosophy is dead. Philosophy has not kept up with modern developments in science, particularly physics. In our quest for knowledge, scientists have become the torchbearers of discovery." Given its limited truth, the statement is so arrogant and dismissive of philosophy that it is difficult to understand or explain the motives behind it.

Stephen Hawking passed away in March 2018. In his posthumous work *Brief Answers to the Big Questions* (Hawking 2018), he formulates, bravely and melancholically, statements such as: "When people ask me if a God has created the universe, I tell them that the question makes no sense. Before the Big Bang, time did not exist; therefore, there was no time when God could create the universe… We are all free to believe whatever we want, and my view is that the simplest explanation is that there is no God. No one created the universe, and no one directs our destiny. That brings me to a profound realization: there is probably no heaven or afterlife. I believe that believing in an afterlife is just wishful thinking. There is no reliable evidence to support it, and it contradicts everything we know about science. I think that when we die, we return to dust."

Although Hawking seems forceful in his thesis, embracing science, he argues that God cannot exist because time itself appeared when the universe emerged from nothing. The idea that time did not exist before that moment is the cornerstone of his argument against the existence of God. But it should be added—and the charismatic scientist would have to accept—that he reaches this conclusion based on the state of science in his time. On the other hand, he is less firm in his thesis when he later comments that "there is probably no heaven and no future life," acknowledging that there is no scientifically reliable evidence in this regard, and that "we are dust and will return to dust."

Hawking arrives at a conviction based on science—that there is no God—but also expresses uncertainty about the existence of heaven and an afterlife, to which he assigns a probability. Reading the rest of the book, we come to recognize that the

British genius places the future of our species in our own hands. For Hawking, heaven and the afterlife are possibilities that future generations may yet attain. Current science does not allow him to accept the existence of a creator. Still, it leaves the future so open that we cannot rule out the possibility that humanity might one day "win" that heaven or overcome time itself. I wonder whether this victory over time—closely related to the myth of immortality—might eventually allow us, through future science, to understand what time truly is, the very concept Hawking uses today to deny the existence of God.

Drawing on the development of modern science since its inception with his admired Newton and its consolidation with his no less admired Einstein, Hawking believes he can glimpse fundamental changes that will be progressively substantiated through new theories and technologies—taking us beyond our planet and beyond ourselves, with the emergence of superintelligent entities created by our own technology. His faith in science is absolute, but not so much in humanity itself. If we do not destroy ourselves before our time—given that we already possess the weapons and resources to do so—then research and technologies in artificial intelligence and the colonization of new worlds open up a future full of hope. Philosophy is not dead; Hawking philosophized using the science available to him at the time.

1.7 Gould's Eureka!

One of my favorite books is *"Full House: The Spread of Excellence from Plato to Darwin"* by Stephen Jay Gould (1997). The well-known paleontologist and science popularizer undertakes, with the erudition that characterizes him, a remarkable study of progress, both human and biological. He criticizes the concept because he believes there is no adequate way to measure it and because it is laden with ideological baggage, preconceptions, and prejudices. His work attempts to demonstrate the absence of positive or negative tendencies in humanity and biological evolution in general. Gould was somewhat surprised that his magnificent essay had not been as successful as he had hoped, since the subject remained in the spotlight and continued to be addressed without regard for his critical insights. For example, we have the study by his Harvard colleague, Steven Pinker, who, in his book *Enlightenment Now: The Case for Reason, Science, Humanism, and Progress* (2018), reasserts the idea of human progress. This is not the place to do so, but it would be fascinating to use Gould's precise conceptualization of progress to reexamine Pinker's claims.

Here, I want to highlight the incredible moment when Gould discovered the key to the subject of progress, which led him to write the book I refer to. He also published technical articles in specialized journals on the topic, but what I find particularly remarkable is his ability to write with precise and accessible prose for all audiences on a subject requiring numbers and statistics—something in which Gould was a true master. He spent years thinking about it. That's understandable, since his professional work as a paleontologist involved studying trends and demonstrating that certain groups of organisms changed some of their characteristics over time.

But something told him that the way progress was being measured was unsatisfactory. And the revelation came to him at a certain point, although it then took him fifteen years to present his results. There were three moments: an intuition, a statistical revelation, and the understanding and explanation of one of the greatest enigmas of American popular culture—the disappearance of the 0.400 batting average in modern baseball.

Indeed, there was an Eureka! moment for Gould. He writes: "All three roots sprang from a common ground, from an intuition born of the most personally exciting mode of thought for any intellectual: that sudden illumination of the light bulb, that eureka! or aha! The moment that turns an old way of seeing things upside down and confers transparency and harmony to what appeared veiled, incomplete, or not yet formulated. I am not referring to an arrogant declaration full of absolutes but to a profound personal experience. These episodes of Eureka! just remove the blindfold from one's eyes and overcome idiosyncratic resistances. Perhaps the rest of the world has always known what one has just discovered. Some eurekas are, however, innovative in a more general sense."

The reflection is invaluable. The idea seemed to Gould like an instantaneous intuition, prompting him to reconsider his convictions about evolutionary progress seriously. Yet intuition, although necessary, was not enough. In science, more is needed. The long desert crossing began from that moment, as he was convinced of his insight. He had to resort first to specific statistical methods—for which he was well-prepared—to give a measurable form to the intuition, and then to study concrete cases in detail, both in human history and biological evolution. Did he solve the matter definitively? I don't think so, but he provided us with fundamental concepts to approach the topic with greater precision.

1.8 Science, Between Iron and Shifting

Although less and less frequently, the expression can still be used: "An interview with the German scientist Sabine Hossenfelder has fallen into my hands"—actually, I read it online, on the occasion of the publication of her book *Lost in Math* (Hossenfelder 2018). The vision Dr. Hossenfelder offers in the interview about current science—especially in the field where she has conducted her research, particle physics—is somewhat disheartening. In her view, there has been no fundamental breakthrough in the "foundations of nature" over the last forty years. No new laws have emerged at that basic level.

Her statement is surprising because, in this field, we have long heard about supposed new laws or explanatory integrations of the fundamental forces of nature—for example, string theory—but the author argues that insufficient attention is paid to a core principle of the scientific method: testing theories against data. Dr. Hossenfelder does not refer to the limits of human knowledge. It's simpler than that. She argues that physicists in the field have forgotten Francis Bacon, and that new data is needed to support the various "theories" circulating around the fundamental forces. She criticizes the widespread attitude among her colleagues that the beauty of equations alone is enough to validate a theory.

Science is not art, and the beauty of its products is not the expected outcome. One must work with data; it is not about choosing the most beautiful equations, but the most appropriate ones to account for observable phenomena. She even suggests that improved or new instrumentation—such as particle accelerators—will not lead us to discover new particles, which she suspects do not exist. We must look elsewhere.

This shows the health of science itself and how, from within, there are always critical voices—on this and any other issue. Dissent has always existed, and I would argue that it must continue to exist. Indeed, challenging nothing less than the latest theories in particle physics is, to say the least, "bravo."

But Francis Bacon's science is far from finished. Hossenfelder is critical of a key aspect of it. There are many other areas where progress is slow. Consider one that is close to me: the field of the origin of life. Could there be a monumental change, as in particle physics, sometime soon? Possibly. Although Hossenfelder argues that quantum mechanics is not a fundamental law—that there is something beneath it, a more fundamental law that we may never reach—I see it as feasible that we will eventually be able to reproduce the origin of life. The field has been stagnant for a long time, lacking new data or better theories supported by them.

Science, however, is dynamic and has made fundamental advances—new theories well supported by new data—in many other fields. In this respect, biology has made giant strides since the discovery of DNA's structure. The avalanche of new data is enormous, so much so that some argue the scientific method no longer necessarily operates from hypothesis formulation to testing; rather, data itself—especially in the era of Big Data (here referred tp as macro data)—if appropriately processed in its entirety, can constitute an explanation without the need for prior hypotheses. Much could be discussed on this point, it's true. But it cannot be denied that while some fields remain stagnant, others—long dormant—explode.

1.9 Not Everything Is Chemistry

Robert Kornberg, Nobel laureate in Chemistry in 2006, declares: "Life is chemistry: nothing more and nothing less." He continues: "The workings of the brain are so poorly understood that we tend to associate them with magical and mystical meanings. But chemically, the brain is a collection of wires and switches… based on a combination of our genetics and experiences… Many people want to attach special meaning to their experiences, such as through religion. But it's chemistry."

Let's take a second reading of Kornberg's statements, but in the opposite key. Imagine that Antonio García, Nobel Prize in Medicine in 2056, declares in his acceptance lecture: "Life is not only chemistry," or later: "The functioning of the brain is much more than a collection of wires and switches." Fifty years after Kornberg's award, García, using the science of his time, argues that we now have sufficient evidence to affirm that life can be explained by a set of new laws showing that it is not only chemistry, but also physics and biology.

In particular, referring to the brain, García points out that by delving into its hidden structures, we have discovered that its organization and functioning—which also obey specific laws—go far beyond being a mere collection of wires and switches. Yet García acknowledges that we are still investigating, and we still do not fully understand or explain either life or the brain. Nonetheless, we have made significant progress since Kornberg's time, and this deeper understanding puts us in a better position to treat them, for example, from a medical perspective.

It will be noted that of Kornberg's two statements—about life and the brain—I have only attributed to García new results based on scientific advances. Kornberg's claim that scientific ignorance leads to magical, mystical, or religious interpretations of life and the brain reflects his personal convictions, and by extension, those of many scientists. But these views have no greater value than those of others who maintain that there is something more than chemistry in life and the brain.

I believe it is a healthy exercise of reason to distinguish between what science demonstrates and what scientists say we can explain. This subtle distinction is more important than it seems—it can become a serious obstacle to scientific development if it leads us into dogmatic positions. For example, if everything is chemistry, I won't study life or the brain from other scientific perspectives.

It is essential to remain vigilant against dogmatism. I can think of no better way to do this than to illustrate, through examples, the difference between science and its products, and what the French philosopher Louis Althusser (1974) described in his introductory philosophy of science courses as "the spontaneous philosophy of the scientist." It is spontaneous for those who, without prior training or instruction in the fundamentals of science and the scientific method, arrive at convictions that may undermine their creativity in science.

1.10 The "Magic" of Laws

Prof. Bartolo Luque is in charge of the mathematical games section of the Spanish edition of Investigación y Ciencia, the Spanish equivalent of *Scientific American*, which also features contributions from Ibero-American scientists. At my request, Prof. Luque sent me his excellent description—published in the July 2019 issue—of a new pattern discovered in prime numbers: the Benford law of the first digit of prime numbers. I almost quoted his explanation verbatim and then consulted my own copy of the issue.

The law can be stated as follows: given a set of data and a digit d (1, 2,..., 9), the probability $P(d)$ that d appears as the first significant digit is given by $P(d) = log(1 + 1/d)$. As Prof. Luque notes, counterintuitively, numbers in many datasets are more likely to begin with 1 than with 2, with 2 than with 3, and so on.

In his description, Prof. Luque presents surprising results on what happens with prime numbers. It's true that for an extensive set of prime numbers (not all, since primes are infinite), any first digit is equally likely—this follows a uniform distribution. But if you take bundles of prime numbers in increasing quantities—say, ten thousand, one hundred thousand, one million, ten million—you find that they follow Benford's law, or more precisely, a generalized version of it. Moreover, as the bundle grows, we transition from generalized Benford's law to a uniform distribution. There is no contradiction with the original proof when a massive bundle of prime numbers is considered.

What "magic" lies in this law of prime numbers? Where do these regularities come from? Many laws have been discovered about prime numbers, and they never fail to surprise. I would almost say we should not be surprised when new laws emerge in this domain. But how is it possible? Why do they exist? Those who apply these numerical laws in various fields rarely ask about their nature. Yet it is an intellectual challenge to understand what these laws owe their existence to.

Returning to Benford's law and prime numbers, I find it equally fascinating that the law holds for specific bundles of primes, but when these bundles become enormously large—as in the case of infinite primes—the law disappears, or rather, a new explanatory law comes into play. In this world of prime numbers, regarded as the building blocks of natural numbers, we find laws at different scales. The law that applies to a million prime numbers (10^6) is not the same as the one that applies twelve orders of magnitude higher (10^{18}).

These results are highly illustrative and can be helpful, by analogy, for understanding how the laws of nature operate. The scale or level of resolution at which we work may yield different "laws." The scale or domain we deal with in physics, chemistry, biology, economics, or sociology may offer laws that work well at their respective levels, but not when we move to others. This is relevant even for understanding why one cannot reduce or assimilate the laws of one scale to those of another. Benford's law explains the first digit well at a given scale, while the uniform distribution does not—although it becomes the appropriate explanation at larger scales.

1.11 The Paths of Science

What we still have to walk on the road of science is almost certainly much greater than what we have already traveled. I turn to the beautiful and existential poem by Antonio Machado, applied to the life of each one of us, to understand what the path of science is—the one we have traveled and the one we still have to travel: "Walker, your footprints are the path and nothing more;/walker, there is no path,/the path is made by walking."

We do not have a marked path ahead of us. However, what we have already learned in science can help guide our next steps. The truth is that there is always a margin of personal initiative to go in a new direction—or, in other words, along a new path that could not be foreseen from the road already traveled. Throughout the history of science, there have been numerous instances when it was dogmatically asserted that the great questions had already been solved, and that the road ahead would follow the path marked by supposedly definitive theories or explanations.

The reality of walking in science is very different. In physics, for example, there are mysteries yet to be elucidated—questions that cannot be explained or that do not fit within current theories—so we must consider developing new ones. These are the new roads ahead of us. We may find ourselves stagnant—even for long periods—moving in circles or following paths that return us to our starting point. But sometimes, those paths are genuinely new, and there is no going back. For instance, in physics, we are currently grappling with turbulence, black holes, high-temperature superconductivity, gravity, the number of dimensions in the universe, and dark matter and dark energy—to name a few areas where new paths will eventually open. In biology, similar challenges exist: the origin of life, the evolution of complexity, the tree of life, brain function, global ecosystem dynamics, and cancer, among others.

The path taken by science, viewed in a panoramic form, resembles an extensive roadmap. The highways and main roads represent the most established knowledge and theories. But I see no reason to rule out that it is on the smaller paths—the roads less traveled, erratic and unpredictable on that great map—where the new roads of science begin. These will, in time, become new highways. It's not that we will dispense with what has already been done; the main roads have brought us to where we are now. They are fundamental, either because they are well-established or because they require revisiting from a new perspective. But with hesitant steps, someone, somewhere, has ventured onto a new path—an explanation or a discovery that had never been considered before. Thus, the path of science restructures itself anew; it is always doing so.

Another stanza of Machado's beautiful poem is especially fitting for those who wish to walk the path of science: "I never pursued glory/nor to leave in the memory/of men my song;/I love the subtle, weightless and gentle worlds/like soap bubbles." The attitude and disposition that should guide us along the path of science is one of interest in understanding and explaining the subtleties of the world, each of its details. Even if it can be achieved, our walk should not be driven by the pursuit of glory or leaving a lasting mark of our accomplishments. That may happen—or not. So, where is the reward? In my opinion, it lies in the act itself, in the emotion and pleasure of discovering something after a persistent search, when the finding or explanation comes to us on one of those paths we have made by walking.

1.12 Of Errors and Prejudices

In my classes on biological evolution, I often refer to the experiments on sexual selection conducted with the fruit fly *Drosophila melanogaster* by Angus John Bateman, published in 1948 in the journal *Heredity*. The results supported Darwin's original thesis, based on his observations of the differential sexual behavior of males and females in various species of fish, birds, reptiles, and mammals: namely, that males try to copulate with as many females as possible. In contrast, females are selective about the males with whom they mate. Bateman showed that in fruit flies, reproductive efficiency and mating success were more variable in males than in females. More importantly, he showed that male fertility increased with the number of matings, whereas female fertility was more closely related to egg production.

These observations formed the Darwin–Bateman paradigm, often used to explain the differential sex roles between females and males: females are more biased toward parental care, and males toward sexual dimorphism (i.e., the secondary sexual characteristics, ornaments, or armaments that males develop in their evolution compared to females).

As is often the case in science, where knowledge must be verified, Bateman's experiments and statistical methods were revisited in two publications that appeared in 2007 and 2012. These papers questioned Bateman's interpretations, finding no statistically significant results to support the differential role of the sexes and their reproductive strategies. Without going into further detail, and as has happened many times in the history of science, this could be considered a scientific error. However, we are dealing with a topic that has been in the spotlight for about 150 years, with multiple confirmatory results. We should therefore be cautious.

Now, let us turn to another issue. The media echoed these findings, suggesting that if Bateman's experiments were not conclusive, we could no longer sustain the idea of a differential role in reproductive success between males and females of any species, including humans. We must remember, in any case, the existence of polygamous species in which the male assumes parental care and the female is the promiscuous one. The media, as I mentioned, gave rise to statements such as: "This dichotomy does not occur either in flies or in people, where studies on sexual behavior reveal that both men and women can feel comfortable in monogamous relationships and can also have equally sporadic ones. But Bateman's thesis, which seemed to legitimize male infidelity and delegitimize female infidelity, had already curdled, dressed with an aura of scientific respectability" (Josep Corbella, "Ni de Marte ni de Venus: el cerebro es unisex," Spanish newspaper *La Vanguardia*, August 28, 2019).

In no way do Darwin's thesis or Bateman's work legitimize or delegitimize anything regarding fidelity in our species. In 2016, another study was published (Janicke et al. 2016), which thoroughly reviewed the matter. The differential role of the sexes in many polygamous species is discussed in detail, and a conclusion is reached. I quote: "Our study confirms the conventional sex roles of polygamous species, in agreement with the pioneering ideas of Darwin, Bateman, and Trivers. Sexual selection research over the past 150 years has not been carried out under false assumptions; however, it remains valid and explains the differences between males and females. Joint efforts of empiricists and theorists are now needed to understand better how and to what extent the environment can modify sex roles beyond the intrinsic differences we have documented here." Science can correct and learn from its mistakes. Prejudices, however accurate they may seem at any given moment, are another matter entirely.

1.13 Science as Openness

I will not deny that what I am about to write is the closest thing to an act of faith—a belief. I maintain that science is our most faithful ally in facing the uncertainties of our existence. And I do so mainly by systematically discarding many other potential allies. In other words, I believe science wins the game if I consider all the tools we can use to support our existence.

Sustaining our existence is much more than recognizing that we would not be where we are now without science. Without considering the rational use of science—or whether it has reached all of humanity and the entire planet—what is clear in my argument is that science is objective and has, by definition, a rational dimension of applicability. But this dimension, as well as its opposite—the non-rational—goes beyond science itself. Other actors external to science can lead us to confusion, making us interpret the non-rational uses of science as products of science itself. This is not the case. That is why I argue that, of all human activities, science is probably the one that best equips us, by its very nature, to sustain our existence.

Once again, to sustain our existence is to consider that, in the hands of science, we are better positioned to address the disturbing challenges we occasionally glimpse: population growth, resource scarcity, aging, and climate change—to name a few truly global issues. We cannot avoid them; we must face them and find solutions. And doing so hand in hand with science is the act of faith—the belief—I bring here for your consideration.

Some thinkers, who are not usually scientists, argue that our species has limits to its knowledge and understanding of the world. Like many other species, we may have cognitive limitations that prevent us from understanding phenomena such as the nature of matter in quantum mechanics—that a particle can be in two places at once, or emerge randomly from empty space—or how consciousness arises from the material substrate of the brain.

If I refer to these examples—among many others whose explanation remains elusive—it is because I am thinking of the infinite character of science. Indeed, science is limitless in its openness to understanding and explanation. This may sound like an act of faith, but we have ample evidence throughout the history of science to support such a belief. How many new worlds have we discovered? What incredible phenomena have we come to explain—precisely through science? Adopting the pessimistic view that everything, including ourselves, has limits is, in my opinion, more of an act of faith than the optimistic view that science does not—and therefore, neither do we. If science is infinite in its capacity to move us toward understanding the laws of the universe, then we go hand in hand with it—and thus, we remain open to the future.

Although it may seem paradoxical to rely on something as changeable as science—given how often its theories evolve—it is still our best ally in opening ourselves to the future. The paradox lies in the contrast with other belief systems whose theses are immovable and to which we cling for meaning. Science is not like that; by definition, it is a dynamic field. It is as mobile in its development as we are. It does not tell us precisely what the future will be, nor does it promise salvation, but it helps us understand that we are part of it—because of how far it has already brought us.

1.14 Mysteries and Limits

We should be cautious about the widespread assertion that there are matters that can never be understood. Let us call such matters "mysteries." Many could be mentioned; here are two of them: the strange behavior of elementary particles—in particular, the fact that some of them can be in two places at the same time—and the mind-body relationship—specifically, how it is possible that, for example, orders of action can be dictated to the body by an entity, the mind, which is apparently immaterial. But if the mind is material, how does it come to be formed? The list of mysteries can become endless; it could even become infinite. Now, I do not believe that we are in a position to maintain that just because the mysteries are there, we have a limit to the capacity to understand them. My thesis is the following: in the face of the infinity of mysteries, we can counterpose the infinity of science.

It is often argued—in an even stronger negationist thesis—that, given the existence of infinite mysteries, we must recognize that there is a limit to our capacity to understand or explain them. To be more precise, some can be solved, but others definitely cannot, because there are insurmountable limits in "us," something like structural impediments, which prevent us from ever being able to understand certain things. We usually resort to what we know about the limits in the cognitive capacities of other species to formulate something similar about our own. The argument goes as follows: just as some species will never be able to understand a prime number, we "must" have some similar limitation concerning another kind of subject. It seems like a humbling cure whereby we must lower our ego as a species and accept that we have some cognitive limitations.

But the argument is fallacious. It is of the type: "All swans are white," or "because we observe cognitive limits in other species, we must also have them." The logic of the white swan collapses as soon as we discover the existence of a black swan. Why shouldn't we be an example of a black swan regarding the limits of knowledge?

Certain facts can indeed contribute to supporting our possible limitations in knowledge. The first is the shortness of our existence. And then also, of course, there is the fact that no one has the capacity to understand or comprehend everything that science, for example, comes to explain. It is likely that, in the past, when science was emerging, you could count on someone who could be said to have mastered all the science of his time. That time has passed. It is hard to imagine anyone, given the continuous expansion of science since the sixteenth century, being able to fully grasp it. But my anti-negationist argument comes precisely from this circumstance. The limit to knowledge must be considered from other perspectives. Science goes beyond what anyone can encompass in their individual life and also beyond what anyone can come to know, given its vastness, which, on the other hand, is constantly growing.

We cannot affirm that any other species has been able to manufacture such a complex product as science, which enables us to understand and comprehend the world around us. And that is an essential factor in awarding us the qualification of "black swan." However, there is even more: although we can accept that there are absolute limits to our capacity to know and explain, we cannot say the same about the capacity for growing knowledge that could lead to the development of products manufactured or built by us. Let us not think only of the most recent developments in artificial intelligence, but also of the importance of creating an arsenal of instruments to know the infinitely small and close, or the infinitely large and distant. Thanks to all this instrumentation, the human species has progressed much further than any new type of progressively more intelligent instrumentation can.

1.15 Macro Data and the Coin Toss

I use the term "massive data" to refer to "Big Data." I do not rule out that we end up using another term—"macro data"—Although I suspect, unfortunately, that what will finally prevail is the term "Big Data."

We may think that what is relevant in Big Data is the availability of algorithms or machines to process vast amounts of data. This is not the most pertinent issue, although algorithms and machines are necessary. What is relevant is the data itself, as making it available on a massive scale enables us to create new perceptions or values about things in the world, organizations, human relations, the economy, and so on.

Here, I would like to bring up something that concerns me. I think that, although Big Data allows us to obtain regularities or correlations that offer us possible answers with a guarantee regarding certain behaviors or questions, as I have just indicated, I harbor doubts as to whether we can derive from them the explanation or deep understanding of what we are looking for when we handle them. In short, I doubt whether we can explain phenomena or things in the world by handling purely and simply macro data. The Big Data advocates are telling us that we should abandon that old Western obsession with causality and stick to the guaranteed correlations of Big Data. This should not be underestimated because a correlation between two variables based on data is not the same as one based on Big Data.

Let us examine an example of causality versus correlation in macro data analysis: the heads or tails of a coin when flipped. What do we mean when we say that a coin, perfect in its design, flipped in a medium with no friction or any other force that compromises its random trajectory until it reaches the ground, has an equal probability of coming up heads or tails? Heads and tails with equal probability indicate a perfect coin when flipped in a medium with these characteristics. Macro data offers a better guarantee than micro data in finding this clue. Indeed, if the number of tosses is not high and we keep a count of them, we may find that we deviate from equiprobability. However, when the number of tosses is astronomical, the equiprobability index provides us with more assurance about the coin's perfection. The macro data will tell us the coin is heads/tails, which is perfect.

Let us make no mistake: here we are dealing with the certainty we acquire about whether something is the perfection or not of the coin when flipping it, because we deal with macro data instead of individual data. That is, what that coin is when flipping it to the point of reaching the scale of the macro data tells you something about its reality, namely: the coin is heads, tails, or heads/tails. I'm not saying it's more likely or less likely, but what it is—it will be heads or tails if it's not perfect, or heads/tails if it is. Macro data can end up favoring the perception of the coin toss without resorting to any causality whatsoever.

What is the causality of the coin flip? If we say something about what the coin is (heads or tails) based on the toss, we can readily appreciate that many factors in the real world will contribute to the toss's outcome. If we know all or most of them—thus, the structure and weight of the coin itself, the force with which it is tossed, the friction it experiences, etc.—and compose them within a theory of the toss, we could find the answer to the following question: why is it going to come out heads, tails, or heads/tails? The macro data of the coin toss will tell you, on the contrary, without resorting to that theory, what will come out—heads, tails, or heads/tails—but not why it will come out. And we can stay there, satisfied, renouncing the causal explanation. The renunciation of causality is a fundamental issue underlying the concept of Big Data. Think about it.

1.16 What, How, and Why

To fully explain the phenomena of life, it is necessary to make inquiries of a very varied nature. These inquiries can be classified into three types, which, in turn, can be grouped into two categories based on their causes. The types—"what," "how," and "why"—correspond to the nature of the phenomena, and the causes are the "proximate" and the "ultimate." We can ask ourselves about food digestion and thus appreciate that it is one thing to know what digestion is, another to understand how it is carried out, and, finally, how it originated. There is an explanation or immediate cause for the "what" and the "how" of digestion, and quite another for the "why" it originated, which would constitute the ultimate cause.

I repeat, once again, that the full explanation of a phenomenon requires answers to the three types of questions, and that not having all of them is equivalent to having a partial understanding of it. This is not the case for digestion, because we can take a very precise definition of it, know what it consists of—which would be its "what"—know in great detail how it is carried out—something like the deep description of the molecular biology of digestion—which would be its "how," and, finally, recognize that digestion is a process so fundamental for the survival of any organism that the "why" of it, how it has come to originate, takes us back to the very moment in which life arises. There is no life without digestion. Given that digestion is closely linked to, or essential to, life itself, and that this process had to take place from its very origin, we could conclude that digestion is an ultimate cause of life, although not the only one.

But we cannot say the same for many other phenomena of life. As can be seen, identifying the proximate and ultimate causes of phenomena can be complicated, and, of course, although both are necessary, they are not equally accessible. Moreover, the difficulties of finding their "what," "how," and "why" depend on the phenomena we consider. We may think that the ultimate causes are more challenging to elucidate than the proximate ones, but this is not entirely true. It depends, as I say, on the phenomenon we are investigating. Let us consider another example—that of the human species—and examine what we can say about its "what," "how," and "why."

If we think about it, we will see the enormous difficulties in knowing "what" the human species is. "How" the human species is also easier to answer, and "why" it exists—its ultimate explanation—also has its difficulties. If we can define a biological species—an issue with a long history of confrontation in evolutionary biology—we can understand that the human species, like any other species, can be defined. It has its characteristics, although we already know how difficult it is to find what characterizes it distinctly from the others. Therefore, "what" the human species is is an approachable issue, but we cannot say we have a universally accepted definition.

The "how" of the human species is a more approachable subject than its "what"; this could be done by describing its details, history, and evolution. The "why" of the human species is not a piece of cake either. How did it come to be? Although not necessarily more complicated than the "why" of any other species, some factors may have contributed to its genesis: brain, fire, language, socialization, and culture, among others. There is broad discussion about which is more relevant or determinant, or what the precise temporal sequence is that has made its emergence possible. I believe that the "what" and the "why" of the human species have a similar level of complication, and, therefore, we continue with a partial understanding of both. We can conclude, thus, first, that proximate and ultimate causes are accessible in some biological phenomena (e.g., digestion) but not in others, and second, that ultimate causes are not necessarily more complicated to establish than proximate causes (e.g., the human species).

References

Althusser L (1974) Philosophy and the spontaneous philosophy of the scientists and other essays. Verso, London

Deutsch D (2011) The beginning of infinity. Penguin Books, London

Gould SJ (1997) Full house: the spread of excellence from Plato to Darwin. Harmony Books, New York

Hawking S (2018) Brief answers to the big questions. Hodder and Stoughton, London

Hawking S, Mlodinow L (2010) The grand design. Bantam Books, New York
Heidegger M (1996) Einleitung in die philosophie. Vittorio Klostermann GmbH, Frankfurt am Main
Hossenfelder S (2018) Lost in math. Basic Civitas Books, New York
Janicke T, Häderer IK, Lajeunesse MJ et al (2016) Darwinian sex roles confirmed across the animal kingdom. Sci Adv 2:e1500983
Monod J (1971) Chance and necessity. Random House, New York
Moya A (2010) Pensar desde la ciencia. Editorial Trotta, Madrid
Nagel T (2012) Mind and cosmos: why the materialist neo-darwinian conception of nature is almost certainly false. Oxford University Press, Oxford
Pinker S (2018) Enlightenment now: the case for reason, science, humanism, and progress. Viking, New York

Chapter 2
Sociology of Science

Abstract In the following 21 essays, I reflect on science as a profoundly human institution, shaped by its internal dynamics and external pressures. I explore the tension between discovery and recognition, the interplay of competition and cooperation, and the role of creativity and will in scientific practice. I critique the compartmentalization of science and the undervaluation of marginal creators, while defending the importance of slow thinking and the need for "wise fools" who persist despite institutional inertia. I argue that science must be governed wisely, with attention to both public and private interests, and that its benefits should be distributed equitably. Through personal anecdotes and philosophical reflections, I advocate for a triumvirate of teaching, dissemination, and reflection. Ultimately, I call for a more inclusive and human scientific culture—one that values openness, critical thinking, and the emotional realities of those who dedicate their lives to understanding the world. These reflections are offered from within a strong commitment to science, aiming to understand its human and institutional dynamics rather than to undermine its epistemic authority.

Keywords Competition · Cooperation · Critical thinking · Discovery · Emotional realities · Openness · Recognition

2.1 The Goodness of Science

We can never thank science enough for its contributions to human welfare. I could draw a simple equivalence—one that is immediately perceptible—between well-being and living well. Ortega y Gasset, in his *Meditación de la técnica* (1982), said that science, in general, and especially technology, a branch of science that translates its findings into concrete applications, allows us to live well. Living well is more than simply surviving; it means going beyond what is strictly necessary for existence. Science has undoubtedly contributed to our survival, but it also goes far beyond that. Science and technology have made—and continue to make—significant contributions to human well-being.

A. Moya, *Reflections on Science and Evolution*,
https://doi.org/10.1007/978-3-032-25682-9_2

It may be objected, as is often mentioned, that science has led to terrible applications, particularly in the development of weapons and their use in wars throughout history. This casts doubt, understandably, on the supposed welfare promoted by science. However, suppose we were to compile a balance sheet with detailed statistics on the impact of scientific and technological advances on the quality of life. In that case, I am convinced the balance would tip in favor of the latter. I would even venture a second reflection: science is probably the institution that most democratically and justly distributes its benefits to all of us, without distinction of origin, class, or condition, promoting individual freedoms like no other.

Please consider this statement in a relative context and assess to what extent other institutions—such as religion, art, or politics—have promoted or hindered, more or less than science, the good life of citizens. Science, by its very definition and method, is universal, and its results must be directed toward that universe—that is, toward us—and, if I may be so bold, toward everything that makes up the rest of the world. An orthodox scientist might argue that science has little to say about "duty," that it is an activity oriented toward understanding and explaining the world, and that while its results may be applied, science itself is not responsible for those applications. That responsibility lies with other institutions.

Allow me to offer a somewhat daring reflection, which I believe is reasonable and takes us beyond science's supposed purity, objectivity, or neutrality. Let us return to the primordial idea of what science is. What ultimately lies behind its exercise—from Galileo and Newton, through Darwin, and up to Einstein—is the desire to understand and explain natural phenomena, a universal aspiration. The fundamental aim of science is not to apply what is known, but to provide clear and systematic knowledge and explanation. That explanation, almost by definition, aspires to universality, and its universality serves all of us. It is a good in itself, even before it is applied by third parties. The openness to knowledge that science offers makes it inherently non-neutral. Its most immediate product—knowledge and explanation—is a good in itself, magnified even further when it contributes, through other channels, to human welfare.

2.2 Explainers and Entrepreneurs

Before my university studies, during what in Spain is called secondary education—that of adolescence—I had the opportunity to nourish my education with literature and its history. The difference between the Spanish Baroque cultural movement styles of the seventeenth century, known as conceptismo and culteranismo, has remained engraved in my mind. The first thing that comes to mind when I recall these two terms is "substance" and "form." Indeed, *conceptistas* such as Quevedo or Gracián prioritized the substance, content, and subtlety of thought. In contrast, the *culteranos*, such as Góngora, gave more importance to form, the beauty of the image, and the senses. I am not going to make a value judgment on which style is better for the development of culture; in fact, both are seductive, although the intensity of seduction varies according to each individual.

While considering the matter of form and substance in my revered Baroque conceptistas and culteranos, the idea occurred to me that both attitudes have persisted throughout history when contemplating, analyzing, interpreting, or constructing the things of the world, and their relationship with science. And then, suddenly, I thought of the founders of Apple, Steve Jobs and Steve Wozniak. Jobs seems to be a clear example of a cultured lover of form and design, while his partner, Wozniak, who developed the programs, would be a conceptualist, more devoted to substance. I have already commented that there is no need to assess which of the two attitudes is culturally more relevant, much less for living; they are simply two visions, two ways of conceiving and interpreting life. Both are necessary and feed back into each other. The example of Apple's founding makes this clear, and its two initiators attest to it, regardless of Jobs's powerful personality and social appeal.

But there is more, and this is where I want to go. I prefer to abandon the dichotomy between substance and form and propose another, clearer one for my purposes: the difference between "explainers" and "entrepreneurs." From the explanations of things in the world provided by science, we can proceed to the applications of science and technology. I do not doubt that technical application is a beautiful thing in itself. Inventors, for example, are primarily entrepreneurs, and civilizations have been transformed and progressed thanks to them. But what precedes entrepreneurship is, in the first instance, an idea or explanation about something that comes from science—and then, in that order, the opportunity for a business idea. I do not argue that one is more relevant than the other, but I know that if the idea is not based on science, no matter how small the explanation, the business idea will have a lower chance of success.

It surprises me—and very much so—that many political bodies and economic corporations in our immediate environment are promoting entrepreneurship without simultaneously placing a similar emphasis on promoting scientific research. The triad is well known: research, development, and innovation. The explanation remains in the field of research, and entrepreneurship includes development and innovation. However, let's not put the cart before the horse, as we risk achieving a low success rate on a large scale. Think of the founding of Apple or Microsoft. In countries with a limited scientific tradition and little or no institutionally consolidated science, there is a danger—or so I sense—of overemphasizing entrepreneurship and leaving the explanation to third countries that have a well-proportioned allocation of public and private resources for research, development, and innovation.

2.3 Discovery vs. Recognition

A fundamental and universal outcome of scientific practice is the discovery of new knowledge. It is essential because science is based on it, and universal in that it applies to everyone who claims to be a scientist. One cannot be considered a scientist without having made a discovery, no matter how minor the contribution might be. This is not exactly a rule, but the very disposition of the scientist—the way they use their method to study the unfathomable nature—leads them to discovery. Indeed, like everything else in life, the quality or relevance of what is discovered makes some scientists different from others in terms of their contribution to the advancement of science. The resources to continue research, the recognition or awards achieved, in strict objectivity, should come in a rigorous order based on the relevance of the findings. However, it is society itself, or the very structure of science, which—external to the scientist—gratifies them. The scientist, internally, must know how to discriminate and remain alert to the distinction between discovery and recognition.

The relationship between the one who investigates nature and nature itself is a relationship of discovery. It may take a long time to arrive, or what is discovered may be a succession of small advances. But what is experienced in that moment is pure joy, probably proportional to the relevance or greatness of what is found. I am not aware of any instrument that can measure it, although endorphins surely play a role, being triggered after solving or discovering something.

Scientific discovery requires creativity, a characteristic that is, in short, a hallmark of our species. We are creative beings; it is part of our essence to be creative, and the absence of such an exercise in our lives somehow curtails or frustrates us. In this respect, scientists are not very different from other groups who, for example, think about or recreate the world rather than scrutinize it.

The scientist cannot be universally understood without the presence of discovery and the joy it evokes. It is a primary and fundamental action. It is essential not to forget that this is what drives the scientist to become a scientist, and not to confuse it with the enjoyment or satisfaction that can be obtained in other ways that are not so essential or fundamental—for example, participation in large consortia, being part of prestigious and well-known teams, having significant research funding, publishing results in leading journals, or gaining media visibility. Discovery and recognition—which can be summarized as two distinct concepts—are two separate concepts. Although both promote enjoyment, only the former should be the objective pursued, as it is intrinsic to science; the latter is incidental, if it comes at all. Although what I have just mentioned may seem obvious, it may surprise us that many scientists, at certain moments in their professional lives, are more motivated by the pursuit of recognition than by the discovery itself. However, the attitude and willingness to continue discovering are clearly evident among the great scientists of all time.

One final thought: reading and studying the great works of the first scientists can help us understand science's commitment to the search for the fundamental laws of nature, regardless of whether the discoveries were large or small. Returning to our origins has salutary effects, helping us not to lose sight of what science should pursue initially: discovery.

2.4 Quantity and Quality of Science

For the development of science, there must be regular funding sources—although there may be others—and, if possible, substantial ones. This cannot be left to improvisation and requires a pact beyond the ideas and programs of the political parties capable of governing a country. Quality is needed, referring to the systematization and regularity of the programs and strategic lines of science, which can be subject to periodic review based on adequately analyzed results. But quantity is also essential. Science managers often tell us that, with the same resources, very different things can be accomplished, and that we should not focus so much on quantity but on the strategic quality of what can be done to promote the development of science. Well, yes and no. I understand the issue of quality, but quantity is significant—especially when considering the countries in our cultural sphere, Ibero-America, which should invest more in science due to its population and economic balance than it has been doing.

When the quantity is meager—even ridiculous—it could be said that the quality sought is offensive to the scientist seeking funding. Quality encompasses controlling and verifying the management of funds granted to scientists, as well as increasing the requirements for forms and documents submitted in funding applications. Why are so many requirements intended to be related to the pursuit of excellence if what will be distributed are crumbs? I propose that someone investigate the optimal relationship between quality and quantity. Quality only improves based on a reasonable quantity, and, of course, it is even laughable to be told that we must seek quality when quantity is strangled. It is simply ridiculous. This analysis should be read as a call for balance within scientific practice, not as a rejection of evaluation or excellence.

Scientists should be protesting effectively when it is clear that quantity has been systematically reduced. Even if we compete for resources, it is unacceptable that we are part of funding agencies and evaluation committees that grant research funds when the resources to be provided are so meager that more than reasonable proposals are ultimately not funded. We must be assertive and decline to participate in decision-making processes when the amount is below a threshold that, as I indicate, could be relevant to us based on indicators of economic growth and the active population.

In my opinion, the scarcity of funding for science in certain countries within our cultural environment is largely due to a lack of tradition in science itself, or the conviction that it does not effectively contribute to the generation of wealth, well-being, and prosperity. It can be accepted that there is no tradition; then it must be created. One must begin. There is always a beginning. The scientists themselves should be in charge of, first, creating quality science and, second, intervening in the administration and the public and private sectors so that the quantity grows, to put the country on the path of its adequate contribution to world science, which corresponds to it in relative terms.

Everything is well studied; we are plagued with comparative reports on economic indicators, population, scientific production, and measures of success in contributing to global science. We know—or can know—if we are not where we should be and, therefore, if we need to inject more quantity. There is no quality, or quality does not appear, if the quantity is inadequate. I hypothesize that our science will have no quality if the quantity is not above a certain threshold. The number of scientists ultimately correlates with the amount of resources; therefore, adequate and sustainable funding can best facilitate the achievement of exceptional milestones, whether through unique discoveries, applications, or innovations of high added value to social welfare.

2.5 Will and Creativity

If memory serves me correctly, the well-known Spanish physician and humanist Gregorio Marañón maintained that true leisure consists in shifting from one type of work to another. He became a member of no fewer than eight Royal Academies of Spain. His intellectual work was pervasive, covering fields as varied as medicine, history, and literature. Reviewing his biography, it becomes clear that to carry out such a task, he had to devote minimal time to activities unrelated to his more flattering studies. In reality—and very probably—a certain iron will for sustained work, even when changing intellectual objectives, must have been at the core of his convictions.

The same seems to have applied to the writer Marcel Proust. After a long recovery from illness and aware of the importance of his literary mission, he wrote in *In Search of Lost Time*: "It is true that I had the intention of returning to live in solitude from the following day, although this time with a purpose. Not even in my own house would I allow them to see me when I was working, because the duty of doing my work imposed itself on being polite and even kind. But I would have the courage to answer those who came to see me or called me, saying that I had an urgent appointment with myself for certain essential things that I had to find out immediately. And yet, as there is little relation between our true self and the other, because of the homonymity and the common body in both, the abnegation that makes us sacrifice the easiest duties, even the pleasures, seems to others to be selfishness." It is worth noting Proust's reflection on how his unwavering will to devote himself to his work could be misinterpreted as selfishness.

Something similar could be said of Charles Darwin, particularly regarding the conviction with which the distinguished English naturalist pursued his work and adapted his daily life to the scientific investigation he had self-assigned. Soon after his voyage around the world on the Beagle, he settled on the outskirts of London, where he carried out systematic and sustained research. It should be noted that he was by no means a person who decided to isolate himself from the world. He devised a strategy to stay informed and connected while maintaining focus on his objectives.

His epistolary correspondence was enormous; he received visits from colleagues and regularly attended London's scientific and academic circles.

On the other hand, it is worth mentioning that Darwin's readings went far beyond his area of expertise. He kept up with new theories and currents of thought from fields far removed from his own. His intellectual appetite seemed insatiable, although he knew how to use insights from other disciplines to nourish his theories. For example, while reflecting on the work of Robert Malthus and his thesis that the human population grows geometrically while subsistence resources grow arithmetically, Darwin arrived at the idea of natural selection.

We are talking about three individuals from different fields and historical contexts who share something in common: a will to pursue an intellectual enterprise against all odds. They knew they could not abandon it, even though they were not entirely convinced that it would ever be completed or socially recognized. Although some of them may have received recognition during their lifetimes, it was not the underlying aspiration that drove their work. On the contrary, it was the will to know, to search, and to explain that shaped the existence and way of life of these three great cultural figures. To generalize, these are the characteristics shared by those committed to artistic or scientific creation.

2.6 Competition and Cooperation in Science

In today's scientific world—now a highly regulated professional activity—the prevailing idea is that any scientific result published, which serves as the primary measure of a researcher's competence, must clearly indicate who contributed what. It has become increasingly common for journals to request detailed authorship contributions: who designed the experiment, who conducted it, who analyzed the data, and who wrote the article. Similarly, when evaluating candidates for positions in research centers—where curricula are increasingly competitive—it must be possible for selection committees to assess individual contributions to key publications.

I wish to state for the record that this has not always been the norm for evaluating scientific quality. It is worth reflecting on alternative models, especially given the challenges associated with assuming that only one evaluative framework is valid. Such rigidity may ultimately hinder scientific development if it fails to identify the most suitable candidates for advancement.

It now seems that cooperation is penalized in favor of competition. While the advantages of collaborative work are often acknowledged, cooperation is rarely prioritized in selection processes. Instead, individual competence is emphasized—excellence, originality, creativity, vision, and leadership are assessed using quantitative criteria that rank candidates. Cooperation is treated as a secondary factor, valuable but subordinate to individual metrics.

I do not intend to propose an ideal formula here. Still, I insist on the need to reflect on alternative ways of conceiving scientific work—particularly cooperation—and to incorporate them appropriately into evaluation processes.

A compelling anecdote illustrates this point: the professional relationship between Godfrey Harold Hardy and John Edensor Littlewood, two renowned mathematicians from the last century at Cambridge. Their cooperative model would likely be negatively evaluated under today's selection criteria. They established a set of "axioms" to define their collaboration, which I summarize briefly:

(a) Axiom 1: "It did not matter whether what they wrote to each other was true or false." That is, there was complete freedom to share even the most speculative ideas.
(b) Axiom 2: "There was no obligation to answer a letter from the other. Nor even any obligation to read it." This is particularly interesting, as it implies a voluntary and non-hierarchical exchange of ideas.
(c) Axiom 3: "They had to make an effort not to think about the same things." This reflects a deliberate strategy to encourage divergent thinking and foster creative convergence.
(d) Axiom 4: "To avoid any discussion, all their scientific publications bore the signature of both of them, regardless of whether one or the other had contributed to their elaboration." Under current standards, this would be considered unethical, especially if the previous axioms were not in place.

However, Hardy and Littlewood operated under a principle of mutual accountability, where their professional relationship was symbiotic rather than parasitic.

2.7 The End of the Scientific Mentality?

Robert L. Laughlin, winner of the Nobel Prize in Physics in 1998, published an interesting book titled *Crimes of Reason*, subtitled *The End of the Scientific Mentality*. The subtitle may cause ambiguity: is it about crimes committed by reason, or about reason itself becoming the victim of a crime? Both interpretations are plausible, as reason can indeed be attacked by reason.

One of Laughlin's reflections—disturbing, yet illuminating—helps clarify the issue: "Inventive people will continue to invent things. What is changing is the economy of the intellect, the traditions by which a person benefits materially from creativity, the conflict between intellectual activities on the one hand and intellectual property laws and national security on the other, the strong incentives to create disposable knowledge, and the increasing cost of locating relevant knowledge in a huge river of garbage. Right under our noses, the Age of Reason is being displaced from its ecological niche by the Knowledge Economy, a term loaded with irony for a time when what is being promoted is the scarcity of knowledge."

The book is dense with insights, but its core argument can be summarized in three points:

(a) Intellectual creation is in decline, hindered by obstacles to the free circulation of ideas—especially those that are most relevant.

(b) Valuable knowledge is increasingly challenging to locate, buried under a flood of irrelevant information.
(c) We are witnessing a subtle but real transition from the Age of Reason to an era of directed reason.

Laughlin, a theoretical physicist deeply engaged with science's big questions, wonders whether much of the new knowledge emerging today is irrelevant, unsubstantial, and disconnected from science's core mission. He argues that this knowledge is shaped by particular interests—controlled by private corporations, protected by intellectual property laws, or restricted by national security concerns. These forces obstruct the free circulation of meaningful scientific ideas.

Moreover, what remains freely accessible is often the least relevant. Laughlin believes we are witnessing the end of the scientific mentality. I am not so pessimistic. The internal dynamics of science may, in the medium term, resist and correct the distortions imposed by external interests.

A notable example is climate change research. Despite strong corporate and national interests that have attempted to suppress or distort key findings, the scientific community—through a decentralized and systematic effort—has managed to create pathways for explaining and publicizing critical advances, both theoretical and applied, on this urgent global issue.

2.8 New Intelligences

It is vital to attract new talent to science through the creation of an international program to promote scientific and technological research on the colonization of new worlds. That colonization, by the very etymology of the term, would be carried out by humans. Reading the posthumous work of Stephen Hawking (2018), I appreciate how closely he aligns with ideas I have also been considering for years. That program is one to promote. However, there is another: the emergence of new intelligences, and the eventual rise of superintelligences that may result from research and development in artificial intelligence.

As a teacher, scientist, and thinker, I always try to convey to students the excitement of science through its discoveries and explanations. It is my own excitement that I try to communicate. We cannot afford not to recruit new scientific talent. Not everyone should become a scientist, of course. However, science instruction is fundamental—we cannot leave science in the hands of a few educated scholars. That would be too dangerous. Therefore, the task is twofold: to teach science and to attract talent. The challenge of colonizing other worlds can be a powerful force for attracting talent, as it taps into our innate curiosity. But another equally compelling issue is research in artificial intelligence.

This is a vast program; many issues demand extraordinary ingenuity under its umbrella. Due to its challenges, it can become a powerful attractor for new generations of brilliant and talented scientists. I have already commented elsewhere in this book on whether the new intelligence that may emerge from artificial intelligence research will be conscious intelligence. To summarize, opinions are sharply divided. Most scientific sectors believe it will be possible to generate a conscious, non-organic superintelligence not based on carbon. The other sector, mainly composed of humanist thinkers and philosophers, is more skeptical about the possibility of machines becoming conscious. However, they do admit that such machines could become more intelligent than we are. It is a considerable debate, which I will not delve into here.

What I wish to emphasize is the proposal that artificial intelligence research should be funded preferentially, with a relatively high percentage of public research expenditure allocated to it, just as I have argued for the colonization of other worlds. The development of new AI technologies is currently being driven by the private sector, which is making the largest investments in this area. I do not argue that this should be prevented in an open and democratic society. However, this implies that superintelligent machines will generate benefits for the private sector that invested in them, or for the specific segment of the population that can afford them. This leads to an ethical reflection: the eventual benefit for a small sector of humanity runs counter to the ideal of general benefit.

I am not trying to limit the public sector—which, ideally, should be composed of the democratic powers of the world united—in terms of what can be researched in artificial intelligence. Indeed, such powers could set moratoriums or red lines not to be crossed, and I would not oppose that. On the other hand, these powers are best positioned to ensure that the products of artificial intelligence research—and those of the program to colonize other worlds—reach all of us and our descendants. Science must serve everyone, as it always should.

2.9 The Triumvirate of Science: Teaching, Disseminating, and Thinking

Popularizing science is very important and praiseworthy, especially now that it seems, little by little, to be becoming a right. However, something is missing from this statement. If all countries truly recognized the right to science, science would not be merely popularized—it would be taught and fundamentally instructed. There is no incompatibility between popularizing and teaching science. Still, we must reflect on the need for both; otherwise, we risk treating science popularization as a mere pastime. There is nothing wrong with that either—what could be better than learning while enjoying it? But that is not the point. The point is that science should occupy, from the outset, its rightful place in the education of citizens. In other words, science can and should be learned through regulated study from childhood.

The concept of popularization has a dimension that, while engaging and capable of attracting attention to answer all kinds of questions, also carries an element of amusement—something like, "Okay, now I understand, and I have the explanation

I was looking for." I am not saying this is unimportant. Many potential readers are drawn to popularization and enjoy scientific explanations, just as they do in other areas of human culture, such as art or literature.

I dare to propose the following thesis: scientific popularization cannot be aimed solely at explanatory amusement. If we learn about science only through popularization—however praiseworthy the professionals behind it may be, including top-tier scientists—we are limiting science's role in society. Popularization should be second-level instruction, built upon the first level: science education in the formative years of children and young people. This is how the right to science should be realized globally. Science is not to be trifled with, although there is probably no better way to learn it than by playing. And we cannot play with it lightly, because humanity's future depends on a society that is deeply educated in science.

And now, the third leg of my reflection involves thinking from a scientific perspective. With it, I configure what I consider to be the logical closure of that more-than-a-resource we call "science"—the scientific triumvirate:

(a) We learn or are instructed in science because it is a fundamental form of knowledge, though complementary to others.
(b) We stay up to date through popularization, which allows citizens to follow the latest developments in science—provided they have received prior instruction. This is how popularization goes beyond mere amusement.
(c) We reflect with and from science on the great questions that always grip us, and for which we can be adequately prepared to answer using the science of the moment.

Note that this triumvirate is designed for non-scientists. I am not expressing a desire to recruit scientists for science—although that is also valid—but rather to make science an integral part of every citizen's intellectual toolkit.

I have repeatedly argued that science accompanies our life journey in a way that differs from other forms of thought, probably providing us with greater spiritual peace. But this does not mean we should abandon it, even if it gives the false impression that it cannot help us with fundamental questions. Science has revealed worlds that were unimaginable just a short time ago. It is a product of the human intellect—probably the one that best suits us. It is ephemeral, in that some achievements replace others, but it is also creative, because it never ceases to renew itself. Science is our most faithful ally.

2.10 Private Science, Public Science

My reflection on science funding will not follow the usual debate over whether the private sector should stop funding science. That would be absurd. It is not only a reality but also increasingly the case that most research today—and even more so in the near future—will likely be funded by the private sector. Still, it is worth considering this form of financing more deeply.

Whenever we discuss the importance of a knowledge-based society and economy, we refer to the increasing role of the private sector in financing scientific research. Large corporations are investing enormous resources in science. Contrary to what one might expect, these resources are often non-refundable, meaning they are allocated to exploring groundbreaking ideas that challenge conventional science and may not yield short-term results. This is because the top executives of these corporations are visionaries who can afford to invest a portion of their profits in "crazy" ideas. However, they are also pragmatic.

We should not be surprised by this model of private sponsorship, as it has always existed. But we must remember that scientific progress comes from scientists, not from the leaders who fund it. Moreover, these leaders do not only support "crazy" ideas. They also fund projects that are more conventional and likely to yield commercial applications. It would be interesting to know what percentage of private-sector science funding is allocated to immediate, marketable products versus exploratory research.

I have not yet mentioned public science funding, which remains substantial in countries that lead in scientific development. Public programs also aim to support societal development and the knowledge economy, sometimes with objectives similar to those of the private sector. And, of course, they also fund exploratory research. So, is there a difference between the two models? Yes—an important one. They are less convergent than they appear, because the wealth they generate is distributed differently.

No public science funding program is designed to create wealth for a specific sector of society. If anything, it promotes equalization and addresses imbalances. When we speak of a knowledge society and economy from the public sector, we mean that the ultimate beneficiaries are all citizens, with the underlying goal of equity. If we could imagine a global state, public funding would be directed to all citizens, not only human beings, but also toward the sustainability of the planet and its ecosystems.

It is misleading to suggest that public and private science funding serve the same interests. If we claim that both benefit society equally, we risk generating confusion. The key question is: Who ultimately benefits from the science funded by each sector? It is often argued that private-sector technologies compete in the market and become widely accessible due to their affordability and cost-effectiveness. But will this always be the case? Will fundamental technologies—such as those related to life extension—be available to everyone? Or only to a few? What kind of society will that create?

2.11 Concentration and Expansion in Science

In 2019, I was on sabbatical leave at Harvard University. Boston and the surrounding area are home to a central nucleus of the world's leading universities and research centers. Young and not-so-young people from all over the world come here to pursue undergraduate, master's, and doctoral studies, or to conduct research as postdocs or senior scholars. Most institutions pride themselves on admitting the best talent, and the selection processes are highly demanding.

One thought that comes to mind regarding this selection is the importance of developing science by concentrating a large number of people from diverse backgrounds in a relatively small geographical area. I have no objection to this process, which I will refer to as "concentration." I do, however, want the selection process for entering top institutions to be sufficiently well-designed to ensure that the best talent—wherever it may be found—is welcomed. Although we are far from achieving this ideal, we are likely closer to it than we were in the past.

Historically, centers of knowledge have always existed, distributed according to the economic, political, social, or cultural strength of the country, nation, or empire in question. The path of concentration for scientific development appears to be the most obvious and is the one that usually attracts the interest of countries and large economic corporations. They do so because they have data on the success of ideas or businesses derived from concentrated talent. Moreover, they have measured the minimum size that such a concentration must reach for the expected return to be significant. This is similar to what happens with the optimal size of cities as wealth-generating centers.

However, there is another path—expansion—which is not designed to generate short-term profitability. If the path of concentration brings together the best minds in science in specific enclaves, the purpose of expansion is to allow science to develop where it has never been or where there is little or no chance of doing so. And why is this important? What need does humanity have to promote this type of action? Does it represent any added value compared to the concentration model in achieving higher scientific excellence?

My argument is twofold. First, science is a human right, and like any other right, public authorities must ensure its effectiveness. Second, expansion can broaden talent recruitment; in fact, it can promote it. We must not forget that talent is not innate—it is primarily developed through adequate education, where science is stimulated, excites us, and gives us the satisfaction of understanding and explaining the world.

I believe that international institutions promoting human rights are well aware of this path. However, countries—particularly those not at the forefront in terms of GDP—should make a double effort: in science education at all levels, and in the creation of research centers. And why not? More developed countries should also contribute resources to promote science in these third countries.

But what is to be gained from all this? First of all, it brings science closer to everyone, but it also accelerates its progress. It adds value to the path of concentration. We need more science to tackle the great challenges facing humanity and the planet, and any effort to find additional paths to concentration must be given serious consideration and thorough study.

2.12 Marginal Creators

I believe the Guggenheim Museum in New York houses the most extensive collection of paintings by the Swedish artist Hilma af Klint. In 1906, she began to develop something new in painting that had never been seen before in abstract art. It is as fascinating to witness the enormous leap into an art detached from the physical world as it is to explore what led the artist to take that step. Her interests in theosophy, spiritualism, and the science of her time seemed decisive in constructing this pictorial world of abstraction and essence.

Additional aspects make af Klint's work even more interesting. One is that her work predates that of artists who were also involved in abstraction and have come to be considered the fathers of abstract art, including Vasily Kandinsky, Kazimir Malevich, and Piet Mondrian. Not much was known about af Klint's artistic activity when they were developing this new movement, and the fact is that she had started down this path at least twenty years earlier. She may have contributed to her lack of recognition as the vanguard of abstract art because she insisted to those close to her that her works—specifically those that initiated the movement—should not be shown until many years later, simply because she considered that the people of her time were not ready for them.

It was not irrelevant to her that she was not known; she believed that her recognition would come later, when the world was prepared to understand her art. I suppose she saw the society of her time as too anchored to the past and tied to conventions, and thus thought it made little sense to be known then. But that does not mean she was indifferent to her anonymity. The hermetic nature of the societies she frequented probably contributed to forging in her the idea—typical of such circles—that there comes a time when society, which is constantly progressing, can "understand" what she wanted to express through her work.

I want to draw a possible parallel between the life and work of highly innovative artists like af Klint and that of scientists who may fall into the same category. Saving the differences in terms of the particularities of their respective crafts and training, a common element underlies both: they go through, assimilate, and learn the art or science of their time to make a great leap.

However, there is another, not-so-positive convergence factor when examining both communities: the uncertain fate of the ideas of some innovators. In af Klint's case, she was reluctant to make her work known during her lifetime. Still, it could well have happened that, even if she had wanted to, her work would have gone unnoticed because she was not part of the artistic circles of her time. Her work might have been relegated to the margins of art history as something minor, upstart, or collateral. Another circumstance that may have affected her is that she was a woman. A similar phenomenon occurred with the rediscovery of genetics. The laws of genetic inheritance were first discovered by Gregor Mendel, a humble monk who worked with peas in his convent garden and published his outstanding results in scientific journals of limited circulation. His laws were "rediscovered" more than thirty years later by prominent scientists working in cutting-edge research centers in the science circuit of the time.

Note the parallel between af Klint and Mendel as two creators working in marginal places, far from the great centers of art and science of their respective eras. One could argue that I have brought to light two similar cases that have survived and received due historical recognition. Af Klint is well known today and is increasingly regarded as a pioneer of abstract art, while Mendel is recognized as the father of genetics. But how many marginal creators have been relegated to absolute obscurity throughout human history? And worse: how much has humanity lost in the process?

2.13 Internalism and Externalism in Science

Within science, it is advisable not to adopt dogmatic attitudes—or at the very least, to be cautious about them. Often, the official theory that dominates a particular discipline is defended by most of those who work under its umbrella. Suppose I say that we must try to avoid dogmatism. In that case, it is because I perceive the thin line that separates the intellectual security provided by the explanatory power of the current theory and the satisfaction or benefits it offers us for being under its protection. Let us refer to this as "externalism."

I admit that in science, we cannot practice relativism, and we do not usually entertain multiple points of view on the same subject. The prevailing theory in a discipline has earned its place for more than enough reasons, and it has probably taken a long time to reach that position, that of being the most accurate explanation available. Achieving that position, I insist, has taken a long time, perhaps even longer than it would take to introduce or develop a new theory. Resisting conservatism to defend the prevailing position against all odds is understandable. However, it is also true that science is a dynamic field.

Accommodating attitudes are as frequent in science as in any other intellectual activity. Schools or currents of art, thought, literature, economics, and so on, have an important sociological component that leads them to either absorb or dissolve innovations or new currents. It is a purely survival-driven act, motivated by multiple factors on the part of those who promote it. Externalism versus internalism.

But science contradicts this tendency by its very essence, even if those who study it agree that its history is replete with such acts. True, but those acts belong to the external world of science, which, although real, should not impose itself on the internal world of science—the realm that ultimately determines scientific progress.

I am not defending a positivist view of science, but rather a critical one. There may be clashes between theories that can explain the same facts at certain times. Reality may be presented to us in a distorted way, and we may find convincing explanations from different theories, which are usually shown to be antagonistic. Since the most common scenario is that one of these theories predates the other in the discipline's history, it is often the one with the strongest external positioning. However, some people demand that this distorted reality be resolved, even if it means challenging the dominant theory. And that is the terrain where science moves and advances.

I am trying to be very general in what I am writing, aiming to reflect the dynamics of science in broad terms. Its safeguard lies in approaching its internality through the purest critical methodology, striving to distance ourselves from both superstructural and sociological externalities, as well as from the accommodating and ideological personal attitudes that bring the scientist closer to dogmatism.

2.14 The Problem of the Compartmentalization of Science

Inventing is not the same as discovering, but there is a relationship between the two, which is closely related to science. Discovery is usually associated with basic science, and invention with applied science. Yet, due to a stereotyped and historically unsupported view, it is often stated that science operates linearly—that the invention of a device follows the discovery of a phenomenon. However, this linearity is not always the case, as inventions sometimes open the door to new discoveries. The dynamics of science are probably best described as the interactive product of its discovery and invention components.

This feedback loop, which may seem obvious, is often overlooked. The organization of science and its funding are typically structured around a clear division between basic and applied research, following the linear model—even though it does not reflect how science actually progresses. You may think, especially as a citizen of a country where science is not a matter of state—although political leaders often claim otherwise during campaigns—that the applied dimension will always be prioritized (i.e., financed), because the social return in the form of benefits is perceived to be greater than that of basic research.

The truth is that this has not been the case, curiously enough, in the most scientifically advanced countries—particularly, but not only, in the USA since World War II. In that country, public science funding has followed the basic-applied linear model, but with substantial allocations to basic research, based on the understanding that private funding would adequately cover the applied sector. This model has proven successful and is often cited as one to emulate.

Still, the issue is more complex. If, as mentioned earlier, the most effective way to develop science is not to compartmentalize its basic and applied dimensions—since this could lead to the forced separation of experiences (i.e., basic science in universities and applied science mainly in industry)—we may end up with two isolated groups. Without the opportunity to feed back into each other, they contribute more slowly to scientific progress.

Therefore, we should evaluate whether progress would be greater if the basic-applied interaction circuit were operational, rather than maintaining a linear model with two separate sectors. There is evidence to support this: large industrial corporations in the USA have adopted a feedback-based model, combining basic and applied research by employing both scientists focused on discovery and engineers focused on development. A relevant analysis would compare the success of this integrated model with that of the compartmentalized one.

If we now consider how science is structured in scientific and academic institutions worldwide, we see that the basic-applied dichotomy runs deep. The challenge is not just about distributing public or private funding, but also about avoiding knowledge compartmentalization. From this perspective, ensuring that science advances rapidly and yields greater social returns is a formidable task.

2.15 Right, and a Lack of Interest in Science

Although it is not part of the scientist's regular training, there is a certain consensus—normative, we could say—on what it means to be a good scientist. It is akin to a set of rules that accompanies or facilitates the process of generating validated scientific knowledge. Sociologists and philosophers of science have addressed this issue, and I believe it is appropriate to draw on some of their reflections here to examine whether something is changing in science, given that a significant portion of the scientific community no longer appears to adhere to these norms. Let us examine a couple of them.

Suppose science is a project aimed at generating knowledge about the world—a knowledge that the scientific community itself must validate. In that case, every citizen should have the opportunity to participate in this project. It is a beautiful edifice under construction, largely anonymously, in the sense that who contributes is less important than the fact that a contribution is being made. We must be careful with the issue of "anonymity," because it is a double-edged sword that may reveal that the project is not being carried out "as it should be." The contribution of specific individuals to science is fundamental; not everyone has the same capacity to make significant contributions. In that sense, we are not all equal—and it's better that way. Just think how beneficial it is for the development of science to have the most educated, cultured, intelligent, and free individuals applying their creativity. And here comes the rub: when it is stated that anyone can contribute to science, we are admitting that "anyone" should be able to contribute. It is not so much the right of "science itself" to count on "anyone," but rather that "anyone" should have the opportunity to be part of "that wonderful project." Well, here we have a first rule that is difficult to apply universally: the right to science.

Another issue of great importance is the lack of interest in doing science. It may seem simple, but it is by no means so—especially in times when science is such an important social institution for humanity, to which people of all kinds, classes, and conditions contribute. Disinterest in the development of science does not mean opposition to it, but rather that it is deprioritized in favor of personal interests. And this, too, is normative. We should be interested in advancing science; everything else should follow naturally, including the differential rewards associated with those who make greater contributions.

As soon as we reflect on this, we realize how difficult it is to uphold. First, because of the entrenched practice of prioritizing personal interest over disinterest. And second, because the institution of science itself is supposed to guarantee the fair distribution of rewards. My impression is that science is one of the best human institutions in terms of justice in the differential distribution of rewards. But let us not deceive ourselves: the evils that afflict any hierarchically organized social institution are always lurking—cronyism, nepotism, and clientelism. The rule here should be to minimize the harmful consequences of these three scourges.

Science has evolved throughout its history, expanding the body of world knowledge. But let us be critical: could this growth have been greater? It's like imagining an ideal model of science that could indicate how much validated knowledge we would expect if the rules I've just mentioned had been operative from the beginning. I fear we are below where we should be, although I take some comfort in thinking that other institutions equally important to humanity are even further from their ideal models.

2.16 Government and Governance of Science

Because of its social and economic importance, science is no longer in the hands of those who create it. Although not guided by the same interests, science is governed by actors outside of it, including those from the public and private sectors. However, in democratic societies, science must be developed for the benefit of humanity; therefore, there is no better way to ensure that science reaches everyone than through a democratic government and society.

It is essential, however, to understand that "government" and "governance" of science are not equivalent. Governance may or may not be adequate for science's development and social reach. Governance implies the sustainable and balanced institutional development of science through the actions of the state, civil society, and the economy. In short, while science is clearly governed today by actors who are not its creators, it is another matter entirely whether it is being properly governed.

Allow me to briefly reflect on the context in which science originated, which may help clarify the thesis I wish to defend. It is often said that modern science was born with Galileo (1564–1642), and that Bacon (1561–1626) and Descartes (1586–1650) provided its philosophical foundations. But we must go further back in time—not so much to find its birth as to understand what made it possible. The promoters of the Renaissance movement—figures such as Leonardo da Vinci (1452–1519) and Copernicus (1473–1543)—illustrated to much of the intelligentsia of their time a new way of relating to and understanding the world. As can be seen from the length of this period, modern science was not born overnight. Certain conditions had to be met, and a profound metaphysical transformation of humanity's relationship with nature and God took time to implement (Burtt 1954).

In my opinion, this historical reflection helps us understand what the future of science may hold. Since I have stated that a profound metaphysical transformation is necessary for its emergence, we must now consider whether new changes are occurring in how we think about the world and relate to it—including society as a whole—that could impact science, especially given its reliance on external actors. These changes must be identified and reflected upon, because they are far-reaching.

Let us consider one such change. Science initially maintained a relationship with divinity. The world it sought to explain was seen as perfect and ordered, and its laws were to be discovered by reading the book of nature. Later, science moved toward the idea that the laws of nature could tell us little or nothing about divinity. Some even argued that divinity was dispensable. Now, however, we are entering a new phase, in which some claim that the world is in our hands and that we can change it—hopefully not destroy it. These are new approaches and a new context that may impact science and how it is practiced, as it is no longer just about understanding reality, but about actively shaping it.

If we add to this the fact that actors with interests govern science—some more legitimate than others—we have a complex situation. Hence, in my opinion, it is crucial to focus not only on the government of science, but on its governance: that is, to ensure its development is sustainable and balanced, even if it depends on third parties.

I can think of no better way to contribute to the governance of science than for its creators—scientists and science thinkers—to become more involved, as a universal collective, in the control and development of science, without excluding the necessary dialogue with other actors.

2.17 Director and Directed in Science

I have always believed that emotion, will, and creativity should fuel the careers of those who want to dedicate themselves to science. Emotion to discover, will to overcome doubts and uncertainties, and creativity to reach discoveries through new or unexplored paths. At the base of these three ingredients lies a fourth, transversal to them all: study and training.

I must admit that there is no universal rule about what someone who wants to devote themselves to science should be trained in; however, I can offer some suggestions. If we follow the path of specialization, we risk limiting creative capacities. I have always thought that the most groundbreaking ideas in a given field often come from contact with, reading about, or training in other fields. However, we must also consider that if one is not aware of the latest developments in their own field, there is a risk of reinventing the wheel. So, a certain formative balance must be struck between the general and the particular. And this balance can be achieved—it is what the best training centers in the world aim for: that students who intend to dedicate themselves to science combine specialized training in their future area of focus with broader exposure to diverse subjects.

The formative phase is followed by the phase that properly introduces the young person interested in science to scientific practice: the doctorate. For this important new phase, one must arrive with saddlebags well-stocked with the ingredients mentioned above. However, a doctorate is not something one begins without the possibility of abandoning it. Without delving into the specifics of doctoral programs in different countries and institutions, I believe it is essential that this phase be structured to facilitate a research career and foster scientists for science. The student must be free to opt out and recognize, as in many other professions, that this may not be their path in life.

What I want to reflect on here is the need to establish conditions so that young people who arrive with saddlebags full of vocation and enthusiasm are not frustrated when they encounter a reality of scientific work that is far removed from authentic scientific practice. And also, why not, those who arrive with less conviction but with curiosity to see what happens if they enter this world. Many studies over the last decade have shown high levels of anxiety and depression among doctoral students, largely due to a lack of appreciation from their supervisors.

This issue must be approached from the perspective of what it means to do science and, as I stated at the beginning, the qualities that must nourish it: emotion, will, creativity, and a balance of general and particular training. There is nothing new under the sun; these are the qualities we observe in the great scientists of all time, including those who developed science in the sixteenth century. Why, then, do we now see so much failure and frustration? Although it is difficult to attribute this to a single cause, I dare propose one that I consider central: the inability to recognize that all members of a research group are engaged in science, with excessive emphasis placed on the supervisor's role, which undermines the student's independence.

The doctoral period should not reduce the student to the status of an apprentice. Like their supervisors, students are scientists practicing science within a team. Not all members indeed have the same conditions for producing science. In my opinion, the fundamental difference between a supervisor and a student lies in the former's more advanced training. However, the remaining discrepancies may well favor the student's contribution to scientific development. What does this way of viewing scientific practice in a group composed of supervisors and students imply? That all of them can contribute to the development of science from their particular conditions.

2.18 Actors and Agents in Science

I often emphasize the importance of returning to the origins of modern science. I do so because I am aware—given what science has achieved as a social institution—not to lose sight, even for a moment, of the fact that the first and foremost element is the discovery of the world's phenomena through systematic investigation. This may sound trivial because it seems obvious, but unfortunately, it is not.

The fundamental element around which all science is centered is not science as an institution, nor the broader framework composed of administrators, managers, and funders. The critical component of science, then and now, is the scientist. However, this may seem self-evident—since, without scientists doing science, science itself ceases to exist—the framework formed by others who support it must not overshadow this reality.

Let me clarify this thesis by classifying all those in the scientific world into two main types: "actors" and "agents." I use this classification to emphasize, once again, the fundamental role of the scientist as an actor. An actor is someone who interprets a role in a play. The scientist interprets the world's play. Without going into further detail, I draw this equivalence to highlight that the actor is the essential element in the play, which is science. Other components may be necessary for the play to proceed, but if the actor is absent, there is no play. The others—the agents—are necessary but insufficient.

An agent is someone who acts with power over another or who is in charge of an agency to manage others' affairs or provide specific services. Administrators, managers, and funders—I may have left out some groups, although I suspect these three include most non-scientists in the world of science—are agents. As such, they are delegated actors; they operate with power granted to them by others or manage others' affairs.

From this definition, it may seem clear and straightforward that agents of science should be at the service of the actors. This is not a relationship of authority or servitude, but rather, in pure logic, "being at the service" or "having a vocation of service" is vital for the monumental framework we call science to function. To make the thesis clear: the actors of science are independent of the agents, while the agents are dependent on the actors—the scientists.

Could this relationship, which seems obvious, be surreptitiously subverted? Could the service vocation be lost on the part of the agents, especially if administrators, managers, and funders end up directly or indirectly controlling the activity of the actors based on the legislation and powers they have been granted?

Unfortunately, this situation can and does occur—more often than we might imagine. And not only in countries with limited resources allocated to science, but also in many of those that provide high levels of funding. I should note, however, that I am not referring so much to the intervention of large-scale agents—such as the political ruling class—but rather to intermediate-scale agents, those who are part of scientific institutions, such as universities or research institutes.

Indeed, research in democratic societies is shaped by major directives from the ruling political class, which, in the final analysis, responds to the demands of the citizens who elect them. But it is not this major player I wish to focus on. Instead, I am concerned with the intermediate agents who, living daily alongside the actors of science, may come to exert excessive control, to the point of curbing scientific activity.

2.19 From Attitudes to Skills

The philosopher Alfred North Whitehead (1861–1947) is known, among other things, for having co-authored the *Principia Mathematica* with the well-known philosopher Bertrand Russell (1872–1970). After an extensive career in mathematics and philosophy at the universities of Cambridge and Imperial College London, he accepted a position as Professor of Philosophy at Harvard University at the age of 63.

Throughout his long academic life, he was deeply committed to promoting higher education. He held various positions and carried out activities related to its advancement, particularly within the London Universities consortium. By re-reading Whitehead and some of his biographical sketches, I have come to appreciate the importance he placed on this mission. We might be surprised today to see the conviction with which he defended the idea that training in classical languages and mathematics was essential for providing students with independence and critical thinking. The history and culture acquired through the study of Latin and Greek, Whitehead argued, had to be complemented by mathematics—a fundamental tool for conceptualization and abstraction.

Having served on education reform committees, it is striking—though perhaps not surprising—how strongly he opposed selection processes and exams as the primary means of evaluating student progress. He told his students that they had already passed at the outset, intending to relieve them of the stress of thinking that the only relevant goal was to absorb specific content and then demonstrate it in exams. Given the importance that examinations and tests have had over time in shaping academic curricula, it is fair to say that Whitehead, and many others before and after him, did not succeed in their efforts to promote a different vision of learning.

But there you have your educational examples. It is the emotion experienced in the process of discovering and understanding that must take precedence. And it is true that in that process, we will encounter new concepts, ideas, and techniques—some of which may be devilishly difficult. However, the emotions we feel when we understand them must be at the heart of the educational experience. The student's attitude must evolve into aptitude—skills that will serve them throughout life.

What Whitehead taught in his classes was not simple. One only has to glance at his magnum opus, *Process and Reality* (1929), to grasp the challenges his students likely faced. Yet he strongly discouraged compulsive note-taking and encouraged students to engage in discussion. He urged them to contribute their own ideas, no matter how outlandish they might be.

There is nothing new under the sun. Many of us recognize that we have had teachers like this—true mentors—regardless of whether the educational system in which we were trained was more or less regulated or constrained by standardized curricula, subjects, and exams. But what did Whitehead, and others like him, want to convey—something that is rarely reflected in our transcripts, yet is perhaps the most valuable tool for navigating the world and shaping it? Undoubtedly, it is the excitement of learning and understanding, and the strengthening of our mental faculties to make that understanding effective. It is not only about entering the classroom with the right attitude, but also about developing life skills.

2.20 The Need for Wise Fools in Science

Shortly before his death, Aldous Huxley published a brief essay titled *"Literature and Science"* (1963), in which he reflected on the paths followed by science and the humanities, and their long history of disagreements. With good judgment, Huxley advocated reconciliation, which he believed should be evident in higher education. I will address this issue again on another occasion.

Here, however, I want to reflect on another theme in Huxley's book regarding the ostentatious differences between science and the humanities. Huxley referred to Victor Hugo's perception of the science of his time, stating that Hugo "believed little in the science of the wise fools." It should be noted that the life of the French writer runs parallel to the nineteenth century (he died in 1885), a time when science as an institution was being consolidated, and Hugo already believed that a large part of the scientific community was made up of wise fools.

Ortega y Gasset (1982) defended the need to promote the formation of educated professionals in the university. The wise and foolish scientists would be an excellent example of uneducated professionals. On the contrary, dilettante humanists would be educated individuals without a profession, yet they contribute to the creation of wealth in society. Despite having detected a significant presence of learned fools in the science of his time, Victor Hugo did not recognize the importance of this group to the advancement of science.

Since I have been reading Huxley's work, I bring here a new reflection—masterful, in my opinion—by Joseph Hillis, an Israeli scientist at the Weizmann Institute, which summarizes very well what this wise fool consists of within the collective of the actors of science. Hillis writes:

"Let's face the facts. Many young people pursue scientific research as a career, but unfortunately, few do so out of genuine passion for the secrets of nature. For the vast majority, it is a job like any other… Moreover, outside academic circles, the extent to which a mediocre researcher can achieve is often unrealized. Except for pure mathematics, almost all scientific research is conducted by teams, and the range of skills among the team members can be extensive and diverse. Undoubtedly, one can perform a respectable job and contribute to the world by being intelligent enough to do what one is tasked with doing… In commerce and industry, some are exceptionally gifted with cunning, ruthlessness, or luck and achieve commensurate success; then there are the vast majority who manage as best they can, and the minority who fall by the wayside. The number of scientists who fail is probably much smaller, and the selection process is proportionately less effective. Indeed, the relative security and stability of a career as a researcher are probably more attractive to the mediocre than the adventure of inquiry is to the brilliant. Where would we get to without this great intellectual proletariat of research?"

Accustomed to thinking about individual talent, perhaps not enough attention has been paid to the fact that this intellectual proletariat—Victor Hugo's wise fool—has been fundamental to the progress of science. Science, indeed, progresses like few other human institutions. Of course, great intellects are responsible for bringing significant advances forward. But these new territories must be consolidated through the very method of science. And that is where the intellectual proletariat comes in.

As Huxley argues in his essay: "One of the outstanding achievements of science is to have developed a method which works almost independently of the intellect of the people who use it... These members of the intellectual proletariat of research are foolish savants and are far less attractive than successful professionals in other fields; however, their method is an appropriate substitute for personal skill."

It is instructive to reflect on this enormous edifice called science, the capacity it has had—and continues to have—to reach where it has reached, and the role played by the wise fools or intellectual proletariat. Such concerns are raised as questions for reflection, not as deterministic predictions about the future of science.

2.21 Thinking Slowly in Science

The current pressure on scientists to publish research results is well known. Those of us who move in this world understand the importance of making a discovery and presenting it to the community as soon as possible. Like any members of a collective, scientists are no strangers to the dual imperative of building a curriculum based on original research and being the first to report their findings. Unfortunately, however, there are many dangers behind these aspirations, because "building a curriculum" or "being the first" are not the fundamental reasons that should underlie or motivate the scientist, however legitimate they may seem. The task of science is—or should be—something else, and it is important not to forget that.

As quixotic as it may seem, here is a warning and a firm proposal to confront the scourge that "publish or perish" has become. Let us not leave in the hands of third parties what is clearly the responsibility of all those who consider themselves scientists: to fight against it. Because what is legitimate—what is consubstantial to science for those who practice it—is the effort to discover and explain the phenomena of nature. The variables of time and opportunity are secondary. When they cease to be so, and instead precede or dominate the scientific process, we enter dangerous waters. Unfortunately, these issues are all too common today, suggesting that something is not working as it should. But I repeat: let us not leave this to third parties, or attribute it to some inevitable effect of the superstructure of science.

Science as an institution today is organized mainly around the curricular merits of those who wish to be part of it. There is nothing inherently wrong with this. In a world of limited resources—such as positions and funding in research centers and universities—it is reasonable that these resources go to the most qualified. But that is the key issue: how do we identify them? This is not an easy question, because the system is perverse and can deceive us. The first priority should be science itself; we should select those with the best intellectual and emotional capacities to create it. But how do we recognize them?

Young researchers are caught in a frantic race to obtain results that can be published, thereby enhancing their credentials and qualifying them for stable positions. Nor are senior scientists exempt—they, too, are under pressure to publish quickly in top-tier journals to gain prestige and secure competitive funding. The former are focused on building their careers; the latter, on maintaining them. Although their motivations differ, they share a common and problematic driver: speed.

Here, I refer to the rush to publish results. You would be surprised to see how many scientific findings published in major journals cannot be replicated. I won't even cite the percentage, but it is significant—and science is based precisely on reproducibility. If this percentage exists, something is wrong with the structure.

Nowhere is it written that science must be done in a hurry—quite the opposite. We must repeatedly verify our findings until we are confident in their consistency and accuracy. And if the system imposes a different dynamic, we must resist it and use the tools at our disposal to return to the right path. A slogan comes to mind: "Think slowly." I do not mean "think little" or "work less." On the contrary, the effort and sustained work that accompany the conviction that one is onto something relevant are good guides. They should accompany this slow thinking that defines the proper way to do science.

References

Burtt EA (1954) The metaphysical foundations of modern science. Doubleday and Company, New York

Hawking S (2018) Brief answers to the big questions. Hodder & Stoughton, London

Huxley A (1963) Literature and science. Harper and Row, New York

Ortega y Gasset J (1982) Meditación de la técnica y otros ensayos sobre ciencia y filosofía. Alianza Editorial, Madrid

Whitehead AN (1929) Process and reality. Macmillan and Company, New York

Chapter 3
Science, Philosophy, and the Humanities

Abstract In this chapter, I reflect on the enduring dialogue between science and other domains of human thought—philosophy, theology, and the arts. I argue that science must not isolate itself but engage in a relationship of equality and mutual enrichment with these traditions. I critique scientific dogmatism and defend the value of metaphysical speculation, intuition, and aesthetic experience. Through essays on geometry, human nature, and the third culture, I explore how philosophical and artistic perspectives can deepen scientific understanding. I also examine the emotional and existential dimensions of scientific inquiry, including the melancholy that arises from confronting the limits of knowledge. Ultimately, I advocate for a humanistic science—one that embraces complexity, fosters critical thinking, and remains open to other ways of knowing. Science, in my view, is not the sole torchbearer of truth but a vital companion in our collective search for meaning. The perspective adopted here is one of dialogue among different domains of knowledge, without reducing one to another or establishing hierarchical supremacy.

Keywords Aesthetic experience · Humanistic science · Intuition · Metaphysics · Scientific dogmatism · Third culture

3.1 Do Not Enter Anyone Who Does Not Know Geometry

This sign appeared on the frontispiece of Plato's Academy, a temple of knowledge in ancient Athens. Why geometry? For Plato, doing geometry meant seeking the perfect numerical rules that govern the universe and everything in it. These rules are not immediately evident, and a confident attitude and disposition to find them is necessary. Such attitudes and dispositions were precisely what one needed to enter the Academy. They did not want to know how prepared you were in specific subjects—as if it were an entrance exam to a contemporary institution of higher education—but rather whether you had a particular attitude toward reasoning, generalization, discovery, and, ultimately, the capacity to think.

A. Moya, *Reflections on Science and Evolution*,
https://doi.org/10.1007/978-3-032-25682-9_3

What a complex issue—and how much it reveals about the quality of education in different societies throughout history. How do we determine who is prepared and whether everyone has had the same prior opportunities to enter the academy? Indeed, how do we detect those who can enter the Academy and deserve to be members of the temple of knowledge? Or do we all deserve it? How has prior education and the environment in which these young people were raised allowed them to access the Academy on equal terms, because they had the opportunity to develop their abilities?

But do we all have similar capabilities? Not really, but I would say that there are enough to begin with. How many are outside who should be inside, and how many are inside who should never have entered? Could there be an academy for everyone? Nothing, a priori, prevents us from thinking that any human being is born with sufficient intellectual power to develop a specific capacity for independent thought. However, the depth of that thought may vary. What a fabulous Academy that would be—with everyone inside, and no one left out. A utopia.

Back in 1930, the philosopher Ortega y Gasset wrote an essay on the mission of the university (Ortega y Gasset 1982). In addition to creating and transmitting knowledge, Ortega saw the university as having the essential objective of training educated professionals. It was to be attended by young people with the disposition to think in a way that is proper to them by nature. To create—what a beautiful word—and to transmit, also lovely, are two fundamental premises for those who live within the university. Nothing is better, when you are in it, than to foster your creativity, a defining feature of the human being. We are a creative species and must promote the development of institutions that encourage it.

But most of those who come to the university enter only to leave some time later. All of them should enter and leave, and most should go, but as professionals and educated people, they should do so with the conviction of returning. But why professionals and educated people? Society needs professionals. That is obvious. The more educated professionals there are, the greater the capacity to transform society and generate wealth. If we speak of a knowledge society, it is in opposition to a society of ignorance. By the way, let us not confuse the society of ignorance with the society of leisure. It is better to know, to know firsthand, and for society to be an amalgam of specialized professionals.

One can also return to the university to continue training and further develop oneself. But beware—let us not forget the second part of the binomial: culture. Many institutions proudly say they are not universities and offer exclusively "state-of-the-art" professionalism. They are also oblivious to the second of Ortega's components: culture, knowledge of the great ideas, and the grand theories. For Ortega, a professional without culture is a barbarian; he has no elements to judge the world in which he lives. Nor is it intended that the citizen who attends university should devote themselves exclusively to developing their culture, since they would be a dilettante. The university's enduring mission is the balanced development of professional and discerning individuals. Discernment is shaped by culture, and the university provides the elements and tools to nurture the spirit. What is achieved with it? Well-educated, cultured, and critical citizens.

3.2 Human Nature

To exemplify the fundamental error of excluding animality from the concept of humanity, I find the lecture by the German philosopher Günther Anders (1902–1992), "Una interpretación del a posteriori," read in Frankfurt am Main in 1930 (Anders 2014), helpful. The title concerns an interpretation of Kant's "a posteriori" knowledge. For Anders, humans gradually settle into the world and adapt to it, integrating and gaining experience, initially starting from its strangeness. Anders attempts to demonstrate that human and animal existence differ precisely because, unlike animals, humans are not embedded in nature but adapt to it by accumulating experience. Animals are in the world "a priori." They do not need to learn it because they are part of it. Although Anders recognizes—in reality, he has no choice if he follows the state of scientific knowledge of his time—that a certain animality does characterize us, particularly during childhood, he quickly hastens to defend the idea that this does not determine or explain man's situation in the world, which he describes as "a posteriori." Hence, according to him, what separates humans from animals is an abyss.

It may give the impression that Anders' reflection today is obsolete, since few would sustain it in light of current scientific knowledge, but this is not the case. A good part of the most belligerent humanism continues to assume the "a priori" assumption—that we and animals are immensely different, even though biological research on animal behavior, since Anders gave that lecture, has led us down paths that make the aforementioned abyss tremble. It is not only a matter of having discovered, to varying degrees, that many other animal species accumulate a specific adaptive experience of learning from a world that is also foreign to them, but also of realizing that this learning must be placed in the context of biological evolution. Everything depends on the extent to which learning is innate and the extent to which it is acquired. Therefore, innateness can explain one hundred percent of the behaviors in some species, as opposed to others, where behaviors are divided between innate and learned. And this is how the abyss between humans and other species narrows.

We are *Homo sapiens sapiens*, but the individuals of other species within our genus, evolutionarily very close to us, such as those of the now-extinct Homo neanderthalensis, 30,000 years ago: how did they perceive the world? Would they have a similar strangeness to what we feel? Let's consider everything we have discovered about their culture, social organization, learning, and so on. Likely, the level of strangeness towards the world of some of their individuals was not as great as the one that Anders raises in his lecture, but it is close to it.

I argue that we must reformulate our foundational assumptions when we realize that our nature is not so far removed from that of other species. My reflection revolves around reconsidering the assumptions we have formulated throughout the history of thought about our ontological and existential status, in light of the corresponding state of scientific knowledge. Neanderthals were discovered in 1856, and studies of paleontological sites yielded evidence of behaviors as familiar to us as burials or the care of the sick. Paleoanthropology has subsequently been joined by paleogenetics, the study of ancient DNA from the remains of species in sites that still contain it and whose sequence can be fully or partially determined. A remarkable milestone was the sequencing of the Neanderthal genome, which verified, among other things, the presence of genes that could have enabled them to speak and, ultimately, live in relatively complex societies. Are we more human than the Neanderthals? We indeed have differences from them, but it is not an abyss that separates us. Without going any further, our genome carries Neanderthal genes and genes from other species of our genus that have yet to be identified—more science, always more science.

3.3 Diversity: A Celebration of Humanity

Much has been written on what unites us because it is common to all of us, and on that which separates us because we are different. Both the former, which we can approximate to identity, and the latter, which we can equate to difference, are loaded terms with a very high and varied semantic burden. Yet, they have been used as bargaining chips to justify some of the greatest atrocities we have displayed as a species. In the name of identity, races, ethnicities, cultures, and nations have been erected. In the name of difference, some races have been segregated from others, some ethnicities from others, some nations from others, and some cultures from others.

In the winter of 2018, I had the opportunity to lecture at the Paideia Foundation in the beautiful city of A Coruña (Galicia, Spain) as part of a cycle entitled *Identities*. The first speaker was the philosopher Víctor Gómez Pin, whose lecture was titled "The Price of Identity." I did not have the opportunity to attend it. Still, knowing the author's philosophy well, with whom I have a long-standing personal and professional relationship, I can assure you that his thesis was along the lines of the social cost of generating identities that ultimately weigh down our coexistence. And indeed, it doesn't seem easy to think of formulating identities for some, if not based on differences concerning others. A sub-collective can only be segregated from a generic collective based on a differential set of properties or characteristics.

For this reason, the different sub-collectives are separated as if they were passing through sieves of different thicknesses. But what is the will behind such a separating action? What drives us to it? Is it a natural tendency in social and cultural evolution? Probably so, and there is a radical need to identify ourselves as belonging to a certain collectivity that provides us with existential protection. The problem comes, however, when the feeling of belonging, out of intimate necessity, is used as a criterion for discrimination, and my belonging to this group is considered something that places me above other groups.

My lecture at the Paideia Foundation, titled after Stephen Jay Gould's book, was on "The Mismeasure of Man" (Gould 1981). I had the opportunity to show how much evil has been done in the name of science when biological identities and differences have been sought, which have led to racism and social segregation. What an ideological burden has existed in many studies in human biology during the nineteenth century and much of the twentieth century on the nonexistent concept of human races. Allow me to elaborate further on this matter and offer some reflection on what I consider a biological element underlying this operation, which seems inescapable to us: the creation of identity in the face of difference. This is the natural dynamic of diversity. As soon as we approach the study of our mere biological nature, our most elementary genetic basis, we will realize—as has been said so many times in recent years following the application of the latest technologies in the study of genomes—that no two beings are identical. This statement is very elementary; it is true, but at the same time, it is very general and of great significance, because, let us not forget, it is valid for any organism or species.

We must rethink group membership as a need for identification to avoid existential helplessness in light of this scientific evidence: my radical and precious uniqueness, a clear product of biological evolution in the first instance, and, secondly, a product of social and cultural learning. Let us set aside our identities and differences between groups and think of ourselves as unique yet wonderfully diverse beings. In the face of identity and difference, let us celebrate diversity as a testament to the richness of humanity.

3.4 Subverting Selfishness

Is it possible to subvert selfishness? It is often said that everything surrounding human action is related to decisions aimed at providing maximum benefit to the individual, including, if necessary, deceiving others to satisfy personal interests. Moreover, with a certain Darwinian whiff, it is argued that this is not only the rational behavior of individuals but that institutions—the economy and politics, for example—which govern and control power in the world, must also be imbued with selfish rationality. Those who can decide in such instances legitimize their actions on the assumption that their behavior reflects the strictest human rationality and that this rationality, in turn, responds to the need to meet the demands of the ego—even if it requires deception. Any reflection that attempts to reconcile personal benefit with the benefit of others is dismissed as a subterfuge, based on the idea that collective benefit, when it exists, is merely a benefit for its components—or, at best, that some individuals act knowing they will gain more than the rest. In short, individual profit appears to be deeply embedded in modern Western society. However, the most troubling aspect of this claim is that we attempt to grant it the status of reason itself: that reason demands selfish behavior as a condition of coexistence.

It was precisely while reflecting on these matters a few years ago—and after reading *Ego* by the German writer Frank Shirrmacher (2014), in which the author explores the consequences of viewing the world through a Darwinian lens, with all its structures in constant competition—that I began to wonder whether selfishness is an absolute determination of the individual or whether it allows for any alternative.

It became a question of overcoming it by introducing a new wedge of humanism, firmly grounded in science, capable of subverting the selfishness that seems to be imposed by the entities that dominate us and compel us to act in our own self-interest. These entities, following Richard Dawkins, are, of course, genes. It is worth drawing a parallel between the selfishness of genes and that of individuals, given that we cannot categorically state whether genes dominate us or whether we are unpredictable to them, even if we behave selfishly. Suppose genes are our secret controllers, or we ourselves—the individual human self—are a genuine evolutionary construct that has emerged and survived in a Darwinian world through selfish behavior.

Let us admit that the world is scarce in resources, for which the respective social structures and organizations compete. Thus, cultures compete, and some impose themselves on others; economies, political parties, and ideologies also engage in competition. Competition seems so embedded in society that we have no choice but to accept that the world is "Darwinized"—an extraordinary reflection, and more than a mere metaphor, by the Spanish philosopher Carlos Castrodeza (2009). Now then: are cultures, economies, parties, etc., superstructures "with their own entity," whose competition is not, in the final analysis, the competition between the individuals that compose them, but rather a competition that takes place among the structures themselves? Let us take the example of the market: who wins in the game of the capitalist market, and in whose arena do investors compete? According to Schirrmacher, the market seems to be a superstructure that is neither reducible nor controllable by individual investors, that has "a life of its own," and is in no way predictable from the investment behaviors of the individuals who compose it. These considerations lead me to wonder whether, in turn, individuals are also unpredictable superstructures for their genes. We can venture the following hypothesis: the selfish behavior of individuals is less determined—and therefore more manageable—if genes do not control us than if they do.

3.5 The Third Culture

In his famous book, *The Two Cultures and the Scientific Revolution (*1959*)*, C. P. Snow concludes that the scientific and literary cultures often overlook each other and sometimes even come into conflict. Having lived with and had friendships with representatives of both, Snow writes: "For I had the permanent feeling of moving between two groups comparable in intelligence, racially identical, not very different in social origin and with more or less equal incomes for their work, who had almost completely ceased to communicate, and who had so little in common in terms of psychological, intellectual and moral climate that instead of going from Burlington House or South Kensington to Chelsea it was as if I had crossed an ocean." The truth is that Snow came to regard such a conclusion as painful because he appreciated, even then, what humanity is losing in its potential achievements and advances of all kinds by the practice of such a partition, which, in his opinion, ends up diminishing creativity. Indeed, that is the net result—not only the actual partition into two forms or traditions of thought associated with intellectual classes—but also the mental effects on all those who are part of them.

It is probably easier to understand what I am referring to if, instead of talking about scientific and literary cultures, we distinguish between the sciences and the arts, including the humanities. At the beginning of my secondary education studies, I was asked, "What are you from? What do you prefer, sciences or letters?" Let us give such a question the importance it deserves, because the world of culture seems to be separated, even today, between sciences and letters. People either say they consider themselves to be of sciences or letters. Or not? I do not know very well why we need to declare or ascribe ourselves, within the collective of people with culture, to one group or another. Still, the truth is that, in my opinion, the only thing we are showing with it is a lack of culture proportional to the part in which we have not been conveniently educated.

Much has happened since the publication of Snow's book in 1959. I am not going to go into the issue of the need for education—which is evident to me—in "the two cultures," and that the modern citizen needs to be well educated in both to be able to conduct themselves critically in a world as complex as the one in which we live. I will have time to do so at another time. Here, I would like to bring up another relevant issue related to the supposed overcoming of the ocean that separates the two cultures, as John Brockman calls it, in terms of access to the "third culture." In 1996, a book edited by Brockman appeared in Spanish with the same title, where he comments: "The third culture brings together those scientists and empirical thinkers who, through their work and literary production, are taking the place of the classical intellectual when it comes to revealing the deepest meaning of our life, rethinking who and what we are." It makes me wonder—and very much so—that the role of the classical intellectual, that of the humanist in Western culture, is almost entirely reserved by Brockman for scientists in the United States of America and Great Britain. In reality, he does nothing more than vindicate the work of great science popularizers and prestigious scientists from around the 1980s, such as Dawkins, Gell-Mann, Gould, Hawking, Pinker, and Penrose, to name a few well-known ones. Are these the intellectuals of the third culture? If one delves into their texts, what one can perceive is not so much an integration with humanism but rather a competition with it—a specific assault claiming, I do not doubt legitimately, access to thought from science.

I claim this in *Pensando desde la ciencia* (Moya 2010). Still, I am very aware of the precedents and the value of the humanities. I do not claim an exclusive or univocal interpretation based on the science of the universe, life, humanity, or society. Why does Brockman ignore essential works by authors of the 1970s, such as Rensch, Lorenz, von Bertalanffy, Monod, Jacob, or Bunge? Why are these not considered better exponents of the third culture? It seems to me, after having read and studied them, that the latter are, and that their proposals are not antagonistic or competitive with the humanist culture. They are probably better exponents of the authentic third culture.

3.6 Let's Learn Philosophy

At the end of September 2017, I was invited to participate in a book defending the teaching of philosophy in secondary education, which, in Spain, has been relegated to an optional subject (Pascual 2017). I presented some of the arguments I extracted from the text I wrote for the book mentioned above. The editors—secondary school teachers—invited me so that, as a scientist, I could give my opinion on the need to teach such a fundamental subject, if I considered it necessary. I certainly do.

Before any specialization, philosophy can orient young people based on their interests or inclinations. Philosophy is the ideal instrument for consolidating a maturing self and, therefore, has a purpose in itself—especially at this formative and pivotal age, which corresponds to the baccalaureate. Contrary to particular or specialized disciplines, philosophy does not begin with concrete instruments, except for the study of the great currents of thought to exercise reason. Philosophy, in the baccalaureate, is an instrument in itself. Studying it is the best antidote to indoctrination. No one who studies philosophy can say that, as a result of its practice, one ends up becoming a dull member of a society that seeks, in its more or less collective unconscious, to lull its citizens to sleep. Because in philosophy, it is true, antagonistic theses are presented, with the notable particularity that all of them are the fruit of adequate, sometimes twisted, exercises of reason. Although philosophical systems must be historically contextualized, reasoning is refined by studying them to the extent that one understands what enables a new system to surpass the other, supposedly. Then, one learns that there is no such thing as overcoming, and what we are left with is a residue that nourishes our way of judging things—especially the world in general. And just at the best time, it can be done: when that young person is shaping their personality.

Philosophy, I repeat, comes before any specialized course. I do not quite understand this formative mania in secondary education that leads to alternative paths—at least in Spain—based on distinctions between the humanistic, the artistic, the scientific, or the technological, and even less so, in fact, during that fundamental age window that is adolescence. The criterion that should prevail in education is the universality of general study. The aim is for students to develop their creative and critical skills and gradually increase their cultural background. The data is not irrelevant in any way, and the teacher is well-seasoned in staging procedures that promote creativity and critical thinking. Hence, the data becomes the happy point of arrival. To do so, they must immerse themselves in a handful of specific disciplines that I would group into four broad categories: philosophy, languages, art and literature, history, science and technology, and mathematics.

What the cultural development mentioned above of the student must be clothed in should not be interpreted as indoctrination or social taming. The basis of all social transformation and evolution lies in promoting and developing creativity and critical thinking in secondary school students. And for this, what better tool than philosophy to put these qualities into practice? How can it be taught as an optional subject in a particular curriculum? It is a contradiction in terms. The universality and the general study to which I referred above demand that philosophy be one of the cornerstones of a young person's education—not so much for its contents (although it is through the study of its history that they are exercised), but for the capacity with which we will be endowed to make judgments, for the exercise of rationality as a whole.

3.7 Science and Intuition of the Things of the World

What characterizes science, above all, is the systematic unveiling of the things of the world—the entities. Science projects, through its method, a beam of light onto these entities to scrutinize their reality. Discovering their essence and what they are may not be possible, but science has the great virtue of examining their inner workings in great detail and clarifying their structure. Intuitions may be better suited to grasp the essence of entities, but intuition does not enable us to master them. Unlike science, essential intuition does not reach the details necessary for such mastery.

That intuition is a direct form of grasping the essence of entities should give us food for thought, because, in some way, we are connected with them; they exist insofar as we see or intuit them. But not because we have a special preeminence over them—rather, because we are united to them by a double principle of materiality and evolution that allows us to be marvelously connected with that which we intend to intuit. Why science, then? The web of entities is not revealed by employing the intuition of their essence. On the other hand, science relies on a procedure of successive approximations and enters into the details.

Without claiming to arrive at any essence, science continually reveals more about the entities that comprise the world. It is a form of knowledge that does not pretend to say what the essence of anything consists of. But, on the contrary, it touches, scrutinizes, and systematically enters into the texture of what is grasped by intuition. Practices of grasping the essence are found in faith or art. The pursuit of the essence of entities is well reflected in the activities of believers and artists. In an exercise of supreme abstraction, they become aware of the world and capture the ultimate essence of what they have intuited—either in a physical work or through inner recollection. On the other hand, it is neither transmissible nor teachable. It belongs to the private domain, and the intimate nature of this realm is one of its most relevant characteristics.

Science is something else. It is alien to solipsism. Science shares, by definition. What is learned through its method is fully reproducible. Science approaches reality and makes it communicable. Science offers us a shared world. With science, we can understand each other reliably; we can always be sure that through its language and discoveries, we have a currency to understand one another. In science, we always know what we are talking about. It would not be scientific if what is communicated were of doubtful interpretation. This is not to say that there are no uncertain or confusing areas in all communication. The point is to clarify the obscurity and make what is established common to all. Compared with other domains of unifying activities, what emerges is that consensus in science differs from consensus in these different practices. The community of agreement in science is profound, deeply rooted, and intrinsic to the field of science. Every participant knows they understand each other and what they are discussing. This cannot be said of other communal practices, because they derive from norms that compel similar behaviors, or they believe they have arrived at similar positions that, in reality, cannot be contrasted. If it is a question, for example, of religious practices, the norms become the standards, but there is no way of contrasting the apprehension of the essences that faith brings, which is of the same nature in everyone. It is even more so in artistic activity, where it is only circumstantial or sporadic that the agents involved—even if they participate in the same orthodoxy—believe they are talking about the same thing when they communicate.

The language of science, including mathematics, is genuinely universal. It is in permanent elaboration and is constructed under the pretension of universalizing secondary access to the nature of entities. I grant that scientific apprehension may be of poorer quality—always approximate, never definitive. However, it possesses the outstanding quality of a shared world, one understood equally. Nevertheless, it is neither a definitive world nor the one we all hope will give ultimate meaning to existence.

3.8 Philosophy Is Not Dead, and Science Does Not Have a Final Explanation

Stephen Hawking and Leonard Mlodinow state on the first page of their book *The Grand Design* (2010) that "… philosophy is dead. Philosophy has not kept up with modern developments in science, particularly physics. In our quest for knowledge, scientists have become the torchbearers of discovery." Given its limited truth, the statement is so arrogant and dismissive of philosophy that it is difficult to understand or explain the motives behind it.

The so-called "third culture" (Brockman 1996) is precisely what is indicated in Hawking and Mlodinow's phrase: that the position occupied by the classical intellectual—whose canonical exponent could be the philosopher—is now replaced by the scientist. Indeed, the world, including the world of ideas, has a Darwinian appearance, and some groups compete with others to hold "the torch of knowledge" and be the ones whose thinking is responsible for providing explanations and meaning in our lives. Although I claim for science the possibility of entering the arena of discussion and reflection on the big questions (Moya 2010), I believe that we do ourselves a disservice if we think that the different forms of knowledge and access to understanding the world are so much in competition that, in the end, some end up replacing others. In reality, they are not.

In a competitive reading of the history of thought, we may get the impression that certain currents, traditions, or schools of interpretation of the great questions of existence end up disappearing in competition with others when, in fact, they resurface at later times with renewed vigor. It is essential to be aware of and think about them, because this exercise is a great source to nourish our creativity. Exclusivism is not the most effective approach to entering the realm of fundamental thought.

Philosophy is far from having distanced itself from science, even if some of its secular currents tend to overlook it. They are close to the critique of reason, science being one of the activities where reason takes precedence. On the other hand, my argument against the arrogance of scientists—not of science—comes from their limited access to the world's final or definitive knowledge. My path to the third culture is more conciliatory with what other traditions and forms of thought, also based on reason, have to tell us about the great questions. It is well worth paying attention to them if we wish to continue fostering a critical sense and a creative spirit. I believe the third culture should be the bridge that really and effectively connects the sciences and the humanities.

On the same introductory page, Hawking and Mlodinow also state: "This book aims to provide answers suggested by recent findings and theoretical progress, leading us to a new picture of the universe and our place in it, very different from the traditional one…" I do not doubt that the picture may be new, but is there only one alternative traditional picture to confront the new one? False—there are many. And, on the other hand, what guarantees do we have, in light of current science, that the new image it gives us of the universe will not be displaced by another one in the more or less immediate future? With almost total certainty, I predict that there will be a new explanation of the universe, because science operates by successive approximations, without discarding, from time to time, revolutionary theories that are unimaginable right now.

3.9 Science and Metaphysics

The well-known British scientist J.B.S. Haldane formulated an interesting reflection on the relationship between science and speculative thinking—let us call it metaphysics—that deserves some thought. He mentioned it in his book Possible Worlds (1927), whose first Spanish edition, dating from 1947, I keep as a treasure in my library. He says, "But I do not think that we know enough about the possible kinds of thought and existence to claim that our metaphysical systems deserve to be taken more seriously than those a thinking mollusk could attain. Such systems seem to aid the imagination rather than explain reality. Nevertheless, it is of fundamental importance that metaphysical speculation continue. Otherwise, we would fall into a stark metaphysical system, like Thomism or materialism, and consider it a matter of course and common sense."

I do not know if Haldane had enough knowledge—probably not, if we consider the state of zoology of his time—about the extraordinary cognitive abilities of mollusks, specifically some species of octopuses belonging to the class Cephalopoda. Today, we know that a certain form of intelligence has evolved in octopuses that converges with the intelligence that evolved in vertebrates (Godfrey-Smith 2016). However, many components of intelligence can be observed and measured in octopus species, just as in vertebrate species. Surprisingly, organisms with a common ancestor dating back at least six hundred million years have independently evolved to these types of intelligence. It is even more astonishing to reflect on how the intelligence mentioned above is achieved through the evolution of nervous systems with very different organizational structures. While vertebrates possess a central nervous system, the nervous system of octopuses—quite different from that of other mollusks—resembles a distributed or parallel system.

To this day, we do not yet know whether some species of octopus have been able to develop a typology of kinds of thought and existence, as humans have, but it is unlikely. Still, if they have, we should not belittle it, just as we should not dismiss our own speculations. As a rigorous scientist, Haldane does not grant speculation any explanatory value in reality. On the other hand, he emphasizes it as an element that promotes imagination. Personally, I find it more fitting to describe speculative thinking as a "metaphor of reality" rather than merely an "aid to the imagination."

Metaphysical systems, like worldviews, help to give meaning to existence for those who formulate them. It seems to me that it is very relevant that we know of them. But Haldane also enlightens us with the dynamic idea that continued speculation is fundamental to escaping the sectarianism of adhering to a definitive metaphysical system. His proposal is very much in line with the always unfinished activity of science itself. To avoid having to confine ourselves, Haldane argues, to Thomism or materialism—to cite two of the many metaphysical systems—it is better to continue thinking speculatively or formulating metaphors of reality. We will gain something from this, not only because we can feed the development of science itself with new ideas, but also because we will be permanently distanced from sectarianism and the temptation to conclude that what certain metaphysical systems affirm becomes evident or common sense. Haldane, how great you are.

3.10 Melancholy of the Man of Science

My book *Pensando desde la ciencia* (Moya 2010) is only 108 pages long. I started writing it in August 1992 and finished it in August 2007—no less than fifteen years later. Then, it took me a couple more years to restructure the text. I ended up realizing that what I had initially written in the form of an ideology—a book of ideas—was more of an essay, because throughout all those years, I went back and forth on the same subject: namely, the relationship of science with thought in general, with academia, and with society. In addition to a few aphorisms and brief reflections that make up chapter four, the first three chapters deal with these relationships. I also realized that I was shaping philosophical thinking from science. Hence the title of the book.

With great boldness, I advanced the idea that the man of science, in his purest essence, is radically melancholic. I commented on something I still maintain: "I still do not know whether to affirm that science is reached by melancholy or that the assumption of science in all its dimensions generates melancholy in those who practice it. Melancholy is a state of mind, and it is generally accepted that one does not reach such a state in response to a particular situation. The individual becomes melancholic when his natural inability to understand or make sense of his existence becomes omnipresent." If we do not have—or do not have recourse to—any other means than science to understand and explain the world and ourselves, we are likely to end up melancholic.

But at the beginning of this paragraph, I was referring to the person of science in the purest essence, which is none other than that which corresponds to how the great scientists of all times—and particularly those who gave birth to science in the Renaissance—have conducted themselves in their inquiry and questioning of the world. Indeed, seeing the positive face of science through its achievements and contributions to well-being, or, more particularly, the professional success or rewards in life that come from discoveries made or theories formulated, accepted, or proven, can provide an important balm for melancholy. But I maintain that such matters are external to the very essence of the person of science—the one who faces the world's knowledge and who, unfailingly, ends up recognizing, with a bittersweet taste, how much remains to be explained and understood through science.

We can be impressed by the spectacular advances in science, but we should not be overwhelmed by them. Let us apply a fair measure, which, to a large extent, will depend on the citizen's training in science, to assess the scientific findings being announced to us adequately. Appropriate training is also needed to sufficiently discriminate between what is communicated as a reliable discovery and what we could call "the promises made to us because of what has now been discovered." These promises for the more or less immediate future often have a greater impact on the citizen than the discovery itself, and it is part of the marketing of science to present both simultaneously.

The ocean we have yet to discover is far larger than the land we stand on. Although it is another boldness on my part, I maintain that the essence of the person of science is close to—or similar to—the essence of the human being in general: the being who lives permanently located on the frontier—as Eugenio Trías (1999) described—the being who always moves on the limit between what she knows and that which she cannot explain, the ineffable. But we are gaining ground on the ocean of ignorance, and the beach on which we always move is a borderland that progressively, but indefinitely, expands. So is science, and so is the scientist who does it.

3.11 The Philosophical Background of Science

Alexandre Koyré (1892–1964) was one of the driving forces behind the modern history of science. I am interested in bringing here a phrase of his that shows how science, despite its internal dynamics—a matter that I will qualify later—is in no way alien to the influence of philosophical conceptions. In his *Études d'histoire de la pensée philosophique*, he comments: "Indeed, I am deeply convinced that the role of this 'philosophical substructure' has been of great importance and that the influence of philosophical conceptions on the development of science has been as great as that of scientific conceptions on the development of philosophy. Numerous examples of this influence could be adduced. One of the best… is provided by the post-Copernican period of science, a period that is commonly agreed to be considered as that of the origins of modern science; I refer to the science that dominated European thought for almost three centuries, roughly speaking, from Galileo to Einstein and Planck or Niels Bohr." (Koyré 1994). Nothing less than the very origin of modern science would be closely linked to important changes in the philosophical positions or horizons of those who promoted it.

We tend to think that science is—or should be—alien to such substructures; indeed, we often believe we should do our best to remove the philosophical scaffolding to remain with the pure product of the achievement or discovery of a new law or theory. In reality, the internal dynamics of science are much richer than the mere substitution of some theories for others or the accumulation of technical discoveries. Science is the product of the persons who build it, and the separation between what constitutes the philosophical substructure and science itself is not as simple as it might seem—despite the supposed obstinacy of the scientist, perhaps due to an excessive positivist attitude, to get rid of what is not proper to science. The amalgamation is so deep that it is difficult to apply any separating surgery. But it is possible to study what each thing is. This is what many historians of science are doing, in the line of research set out by Koyré in his time, trying to show us: (a) how scientific thought is never entirely separated from philosophical thought; (b) that especially the most stellar moments of science, when substantial changes occur in it, are determined by changes in philosophical conceptions; and (c) that scientific thought is always associated with a set of very fundamental ideas or principles of a philosophical nature.

As I have already mentioned, it is the task of historians or philosophers of science to uncover or reveal what these fundamental frameworks comprise. My thesis is that scientists are not often aware, most of the time, that they are working under certain fundamental assumptions. Even if one is the promoter of a significant change in science, one must have undergone a previous philosophical shift. To summarize it in a metaphor or an image valuable for science, I would say that different conceptions or visions of the world will be determinants for formulating new scientific theories.

In 2013, I wrote the prologue for the posthumous work of my admired philosopher and friend Carlos Castrodeza, *El flujo de la historia y el sentido de la vida: la retórica irresistible de la selección natural* (Castrodeza 2013). I clarified that the metaphysical substratum underlies Darwin's theory of natural selection and Castrodeza's outstanding final thesis, which aligns perfectly with Koyré. There, I wrote: "Castrodeza, while being a faithful ally of science, insofar as it is a sublime rational activity, unveiling the ineffable, concludes that science is not alien to an underlying metaphysics. No human activity is alien to the ideological context of the corresponding epoch, which permeates or impregnates any practice, however intellectual, rational, or scientific it may seem to us. Castrodeza concludes, after a detailed study of Darwin's thought and his time, that the naturalist's work can be perfectly contextualized in his time and that his formulations, including the concept of natural selection, fit naturally well into the thinking of Victorian society."

3.12 Things Are in Place, but with Perspective

In 1924, Edwin Arthur Burtt published an influential book on the metaphysical foundations of modern science (Burtt 1954). Since then, there has been considerable intellectual production on the birth of modern science, with Copernicus, Kepler, Galileo, and Newton, among others, being among the most notable. Medieval science—if we can speak of science in that period—is nourished by the philosophy of Aristotle and Judeo-Christian theology, which succinctly puts humanity at the center of the universe. Modern science is based on the knowledge of the world and the laws that govern it. With a subtle questioning of the most straightforward explanations—heliocentrism—to account for the movement of the planets (Copernicus), the assumption that the universe is governed by mathematical laws (Kepler), or the distinction between primary qualities that characterize things and secondary qualities that man perceives about them (Galileo), we arrive at a new metaphysical position that affirms that nature does not serve humans, but rather that humans are just another element of it.

With the further development of science—although it took a couple of centuries more—Darwin finally demonstrated that the human species is a product of biological evolution, like any other, and even that, just as we have come to be, we might not have been. The worldview emerging with the evolution of science seems to culminate in a kind of "putting things in their place." Our planet is not the center of the universe, nor is it because we are the ones who inhabit it.

In our major demonstrations to the rest of society, we scientists tend to adopt this type of cosmological position, confronting medieval cosmology, implying that modern science has definitively overcome terracentrism and anthropocentrism. But is this a definitive position? Do we feel committed to it, and do we not see any alternative possibilities? Yes, this is the opinion held by the majority, and almost without a doubt, we could say that the revolution in science, which began with Copernicus, Kepler, Galileo, and Newton and culminated with Darwin, constitutes the metaphysical framework of present-day science. Can we expect any fundamental change in this metaphysics? I think it is worth reflecting on this because, in line with what has been central in my thinking about thinking from science, it is always worth considering the possibility of some relevant transformation in the metaphysics that underlies current science—precisely because of the evolution of science itself, as it throws us novelties and results that could shake our fundamental assumptions.

With science, we cannot state categorically that "things are definitely in place." We cannot do so because we may encounter an unexpected turn around the corner, with a discovery or a theory that could lead us to new fundamental insights. Without stating that there is no other explanation, some physicists have come to argue that the fundamental constants governing the known universe—if different—would make life or intelligent life impossible. This is known as the "anthropic principle." And from the field of biology itself—although also closely linked to certain theses maintained by physicists who study complexity—voices are heard about the inevitability of biological complexity here and there in an expanding universe.

The most recent science, taking into account these considerations, returns once again—though with many pertinent nuances and precisions—to what was sustained in medieval philosophy and Judeo-Christian theology. Things are in their place, but we cannot exclude the possibility that we may enter into a new cosmology, precisely provided by what could be fundamental discoveries of the most current science. Worldviews can become shifting sands with the advancement of science, although the metaphysical systems created from it can be maintained for centuries.

3.13 Science and Art

Every weekday morning, I walk from my apartment in Fresh Pond to Alewife, Boston's Red Line subway station, get off at Harvard Square, and then walk to my office across Harvard Yard. In the afternoon, I go the same way back, but in the opposite direction. It takes me no more than an hour to walk to and from. It's a routine that I take advantage of to enrich with reading; I don't see anything better to do than to read during the subway stretches, as do many other people. The book in my hands, which I have only read at times, is a series of lectures by the Argentine writer and mathematician Guillermo Martínez on the interest and presence of mathematics—and science in general—in Borges' work (Martínez 2012). I can see how Borges' intellectual concerns brought him closer to science, although I do not know to what depth. The writer likely had a philosophical substratum of a Platonic nature, which drew him closer to science, particularly mathematics. This is why it is not surprising to observe the parallelism between the worlds that the brilliant writer constructs and those that science seeks to explain.

In a recent piece of writing (Moya 2018), a collaboration for an issue of the journal *Arbor* entitled "Humanities and Scientific Thought. Literature and Image in Front of Science," edited by Prof. Manuel González de Ávila from the University of Salamanca, I had the opportunity to reflect on the same questions that Guillermo Martínez narrates in his detailed study on the mathematical components of Borges' world. That is why I reaffirm the wisdom of bringing the sciences, humanities, and art closer together.

My fundamental thesis is presented at the beginning of this work in an introduction titled "From Intuition to Craft." Suppose we abstract from what science represents at present—something like a fundamental social structure that contributes to the economy through knowledge—and instead focus on the idea of the creative scientist. We will soon be able to appreciate the enormous correspondence this figure has with the artist. The artist, too—probably on a smaller scale—can contribute to the economy from the perspective of art as a business. I do not object to either approach: science and art as wealth generators. However, the convergence I want to bring up here between the artist and the scientist is another and more fundamental and intimate one: that of the satisfaction and—why not admit it—the spiritual pleasure obtained through the process of creation itself.

Remember that the hedonist doctrine holds that pleasure is the supreme good. Although there are different interpretations of hedonism, I am referring to the version of spiritual pleasure that individuals experience when engaging in creative activities. In that article, I wrote: "In the end, whether their respective worlds are real or symbolic, and regardless of whether or not through science we have access to the only possible real world—a matter of eternal debate—science and the arts converge, in my opinion, in a clear and consubstantial hedonism in their practitioners: scientists and artists. Again, whether real or symbolic, worlds are accessed in the first instance through intuition, the idea, and the will to understand them."

Whether the world the scientist constructs is real or symbolic, the creative process is closely related to the figurative worlds of the artist. Moreover, the truth is that scientists create figurative worlds in the current state of science; for example, we have virtual reality. Art—or rather, some artists—have envisioned the real world in multiple ways, thus recreating it. Under these considerations, we observe an important convergence of objectives between the two professions. But the greatest of them probably lies in the undeniable satisfaction that the creative act produces in those who perform it.

We can also consider the benefits that the creative acts of scientists and artists bring to others on a social scale, which brings us back to the previous thesis on the utility of science and art. Still, I think of this variant of hedonism more as a derivative of the essential hedonism that individual spiritual creativity presupposes.

3.14 Dogmatist or Skeptic?

Against dogmatism, we must be forewarned. A significant portion of the concepts used throughout the history of thought have undergone changes from their original definitions to their current meanings, including the possibility of coexisting with multiple interpretations. And that is what happens with the term "dogmatism." Dogmatism is a philosophical doctrine opposed to "skepticism," which affirms that the human mind can know the truth. However, it also conveys a different meaning—undoubtedly the best known and most widely used—which refers to the attitude of a person who does not allow their statements, opinions, and ideas to be discussed. I want to reflect on the relationship between science and dogmatism, considering both their meanings.

Let us first deal with dogmatism as a philosophical doctrine. It isn't very easy for science to affirm such a thesis. We can determine whether the human mind can discern the truth through scientific inquiry. Indeed, we can see the world, things, entities, and the universe through science. We explain them in a continuous and systematic approach. It is probable that, with the success of such explanations, we can affirm that science—which is the product of the activity of our mind—supports the thesis of dogmatism, which is that, collectively, it is capable of knowing the truth.

Now, let us think this over at some length. Let us suppose that the bag of truths is there and that we progressively draw out balls and more balls, each of which is true. But that bag: is it not infinite? Are there not infinite truths? And what guarantee do we have of being able to access all of them? Because they are endless, we can think the bag of truth will never be emptied. We are almost leaning towards a certain skepticism, contrary to dogmatism. In its weakest form, we could say that since there is an infinite number of truths, we will not be able to deal with all of them. But the hardest variant of skepticism is that some truths are not—and will never be—in the bag from which to draw them, because they are truths unattainable by our mind, and science, which is a product of it, will not be able to discover them.

So, which doctrine do we opt for in science: dogmatism or skepticism? I think neither. It is essential to qualify this, because I would be a dogmatist—in the second meaning of the term that I wanted to consider here—if I were to say that I opt for science for both dogmatism and skepticism. Nothing can be said in this respect that science could sustain or demonstrate. However, I can "believe" in the unlimited capacity of our mind to know the truth, just as I can "doubt" that same capacity. As I present it, I want to imply that science's faith in both doctrines is right in the middle. Or is it?

Do we have any evidence that science will contribute to the extraction of the ultimate truth from the sack of truths, or even go so far as to consider that all truth will be in that sack? I can only think, once again, of betting on dogmatism based on the way science has gone so far—unveiling new truths and having faith that others will be discovered, which, to all appearances, might seem implausible right now. But such a belief is just that, with a certain dose of reason based on the achievements of science. The leap to dogmatism would occur when I formulate the statement—which I do not make, but many other scientists do—that truth is accessible to be unveiled by the human mind.

3.15 Counterfactuals, History, and Science

There is a particular tradition among historians of creating fictitious stories, but seasoned with a certain verisimilitude. It involves constructing a tale that considers what would have happened if a specific character had made a different decision in the past than the one he actually made, or if an event—also in the past—had had an effect contrary to the one it had. Against the facts as they occurred and the consequences they had, "counterfactual" stories are built by considering, as I say, personal decisions or outcomes that differ from what actually happened, and creating a new story that is as plausible as possible based on such changes.

History is what it has been—unique, one could say. However, I will not discuss here whether we can reconstruct it with the precision and objectivity that would be ideal, among other things, because of the well-known claim that history is written by those who win. Be that as it may, and since I am not very much in favor of relativism, I will consider that there is only one chain of events that has been determinant for phenomenon X (for example, any civilization), even if we do not know for sure exactly how it unfolded or the causal relationships between its elements. And of course, if there is only one, it is not an insufficient exercise to construct plausible parallel stories—if only for the pure pleasure of imagining how the history of that phenomenon could have been if the chain of events that promoted it had changed at specific moments, if decisions had been taken that were not taken, or if events had occurred that did not take place.

Seen in this way, and considering that each phenomenon has "its" history, the number of counterfactual stories would be infinite. However, only a limited number of them would have the category of plausible counterfactual stories—namely, those that could have occurred, generating changes to real history that would still seem relatively feasible to us. But let us refine this reflection on feasibility a bit more. Of all the possible counterfactuals of X, the most viable are those that we have been able to corroborate as having existed in phenomena similar to X (for example, other civilizations) but corresponding to another time and place in the history of humanity. That is to say, if the phenomenon X that interests me is similar to another Y. Still, the first one has had a chain of events in its evolution different from the second one, then Y is a good counterfactual of X. We can use our imagination to think what the history of X would have been like if this or that had changed in its occurrence and to what extent those changes resemble those observed in Y. This is what gives verisimilitude to the counterfactual of X: verifying that there is a Y where it seems that the changes in the events that did not take place in X did take place.

Why do I consider this issue of history and counterfactuality so important? Contrary to what is often held, there is more science than meets the eye in history itself. The route followed by X, if it could be accurately and objectively detailed, could be compared with routes followed by other X-like phenomena, where the corresponding details—also accurately and objectively available—could show why they ultimately became different stories. However, the corollary that we can derive from analyzing all these phenomena, which are counterfactual to each other, is that we might be in a good position to predict the future of a current phenomenon with X-like characteristics.

I have given considerable weight to the history of humanity. However, something similar—and where we probably have more possibilities of precision and objectivity for reconstructing the corresponding historical phenomena—is the history of life and biological diversity. Since there are clear parallels between the two, I suggest that we accompany the study of human history with that of the history of biological diversity, as it is replete with counterfactuals that can help us make more accurate predictions about humanity. I am not resorting to any naturalization. I only ask that we consider the parallels between human evolution, its history, and biological evolution.

3.16 Identity, Metaphysics, and Science

Personal identity has a long tradition of thought and forms part of one of the fundamental questions of metaphysics: Why am I the same as I was in the past if I have changed so much, both materially and psychologically? The reflection I am going to make here is not going to take up how different philosophical traditions have approached the matter, but rather to present specific considerations that, in the light of science, can be made on such a thorny and paradoxical matter. Indeed, how is it possible for me to maintain that I am now precisely the same as I was long ago? I have changed materially. I have been renewing myself, since all the constituents of my body, in a sort of happy dynamic equilibrium, have been replaced. Where there was a concrete X of my material, everything has been replaced by another similar X, which does the same thing, or almost the same thing, and by another, uninterruptedly, to carry out its secular task.

But our appearance is not the same from the beginning; far from it. It is worth considering whether it is not only that we are not the same, but that we are not the same before and after. I speak of a material replacement as if we were made from the beginning in one piece. Not at all. Our material being is complex in time, a characteristic shared by all living beings, and we are no exception. We all have the peculiarity that we are becoming; we go through stages of development, and part of our materiality remains unchanged in its function, while its components are properly replaced, but another part is new. It is being added following a program, and what is added also undergoes the replacement process in its components to maintain the latest functions. Despite this, and returning to humans, given that we are aware of such phenomena, each one of us continues to manifest that "I am still me," that I was also the one who was born one day. In any case, it is interesting to know how far I can trace this idea of myself, a secular debate with many religious, ideological, and political edges. In any case, I see the photographs, videos, and memories of my life that others claim are mine, at least when I was unaware of being myself, and then all that, but with full awareness on my part. And, of course, I am astonished to witness a continuous replacement of my material components and a transformation of myself.

It seems that we are endowed with a kind of observatory from which we carry out the panoramic vision of our life and impregnate it all with our untainted selves. But since science finds no place to accommodate any soul, since this observatory, if it exists, cannot be something alien to the rest of our permanently replaced and changing materiality, how is it possible to have such a stronghold that gives us the mark of the self, of our singular identity? In reality, although much remains to be discovered about the relationship between mental and psychological states and the material bases of the cerebral processes that generate them—something about which there is an enormous amount of scientific and philosophical literature—what is certain is that consciousness, in particular the consciousness of the singular self, is inscribed, like the rest of the cerebral processes, in the program of our development. The self and identity are reached, and once they appear, they are installed as the lighthouse that illuminates our existence. Not all beings enjoy such a quality, if that is the right word. Some are aware of belonging to a group or species, and others are aware of their own existence. Still, the dubious privilege of being aware of your singular identity seems less common. Let us not commit the anthropocentric arrogance of claiming identity as unique to our species. Science has taken—and is taking—many steps in this area. What is certain is that it has appeared, albeit sporadically. What power does life have to, through its different evolutionary avenues, endow us with identity?

3.17 Rethinking from Science

I was thinking from science. The indication I make that what I am doing is thinking "from" science may give the impression that it is a partial view of thinking, insofar as I am developing any topic from a particular perspective—namely, what science says about it. Although it might seem that there are non-partial ways of thinking, i.e., capable of accessing absolute knowledge of the world's reality, I sense that this is not so, that there is no such universal thinking. The world is always much broader and more **complex** than we can imagine, and we access it from regional thoughts, each one attending to the different registers that the world contains. That is why it is convenient to place science and other forms of thought in their proper place. This is not relativism, but rather the recognition that the complex and infinite reality of the world can be understood through forms of knowledge tailored to its various aspects or registers.

Let's do a brief reflection exercise and divide our thoughts into two parts. One would correspond to thinking from a scientific perspective, and the other to thinking outside of science. But because of what has been said above, we must also consider the reality of the world, divisible into two parts: that which is susceptible to being thought from science and that which is felt from other forms of thought. It is convenient to remember that although the world's reality is one, its spheres are so many that "there is land for all." Moreover, we have not yet finished defining the whole field of the reality of the world that appears to us more pastel than the reality to be incorporated, but what form of thought corresponds to it?

As can be understood from what has been said above, my "thinking from science" is not a "scientistic" way of thinking. "Scientism" refers to the claim that science, by virtue of its method, constitutes the only legitimate and most valuable form of human knowledge, to the detriment of other epistemic domains, and should not be conflated with science itself as a rigorous and indispensable enterprise. Moreover, to the extent that science measures and quantifies, anything that does not fall within that scope is outside of what science can know and is therefore considered to be of little relevance because it is unreliable.

Since scientific knowledge can employ its concepts and instruments to measure the things of the world with certainty, it is presented to us as the indisputable and safe form of knowledge; that is what "scientism" maintains. It is currently prone to dogmatism, as it overrides other established forms of world knowledge in the history of human thought. These different forms, whether they are called philosophy, art, or theology, not only provide explanations of the world but have also been decisive for the gestation of movements and advances in humanity until we find ourselves—for better or for worse, it must also be said—where we are today.

Therefore, my "thinking from science" is compatible with thinking "from outside science" because the world's reality is so vast that there is ample ground for scientific and non-scientific thinking. However, the frontiers that delimit reality are far from immovable, although neither is reality itself, because all forms of thought lead us to discover new folds of reality. It would be as dogmatic to consider that the boundaries of scientific versus non-scientific thought are immovable as it would be to think that reality is fully known and divided into these two parts. Both the knowledge of the world's reality and its very breadth are pure quicksand.

All forms of knowledge or apprehension of the world outside science have their evolution, and to the extent that they change and improve their capacity for thought, they are applied to new areas of reality. However, science also plays a role in this dynamic game. On the other hand, although it is aware of its limitations, it is also cognizant, given its history, that it has no limits. Through its unquestionable capacity for renewal, science broadens the spheres of reality of the world where it is applied.

3.18 Science and the Apple in the Fruit Bowl

The philosopher Markus Gabriel is one of the current defenders of realism. We should be cautious with the idea that there is only one reality and only one way of approaching it: using scientific knowledge. Gabriel introduces an interesting concept for understanding what reality is composed of, which he calls the "field of meaning" (Gabriel 2017).

I will try to explain it with an example. Let us consider a single apple placed in a fruit bowl. There are many fields of meaning in which this object appears. Thus, we have the field of meaning where the study of the ultimate atomic constitution of things, including the apple, is located. A second one could be the field of meaning constituted by biological processes, particularly the one that has given rise to that fruit. A third could be the process that has taken the fruit to the fruit bowl, from the time we bought it, along with others, until only one piece remained because we ate the rest. Another one could be the field of my appetite right now for that apple, and, finally, the field of meaning that could be that the apple in the fruit bowl is only a dream on my part. The reader can think of other similar fields of meaning.

As can be seen from the list I provide of possible fields of meaning associated with the apple in the fruit bowl, we can easily deduce that the number of fields could be ideally infinite. It is very interesting to consider this circumstance because it reveals that reality is much more complex than it initially appears. The apple exists, although it should be said that it is not a single fact or object but innumerable ones, depending on its location in the different fields of meaning.

We might be tempted to think that of the five examples I have mentioned about the object "apple," only the first one corresponds to the actual apple, that of its atomic structure and the ultimate physical laws that can explain it. For the sake of the autonomy of biological explanation from physics, we could even consider that the process of formation of that fruit follows laws proper to the development of a biological entity, and that it also makes sense to speak of the object "apple." But the other three fields of meaning would not give reality to the apple, namely: how it is that there is only one apple in the fruit bowl, my appetite for it, and, of course, the most complicated of all: that what I see—the apple in the fruit bowl—has been a dream.

Gabriel claims that reality is more complex than it seems. True, he argues that science can tell us a lot about the object that is the apple. Still, its reality does not end there, because every time we place it in other fields of meaning, the object "apple" has a differentiated existence in each of them. And the apple, in its atomic and biological composition, is no more an object than how it came to be the only piece in the fruit bowl, my appetite for it, or having dreamt it. All these objects or facts related to the apple "are real," and there is, for Gabriel, no greater relevance of some than of others.

It may even be surprising to consider "the dream of the apple in the fruit bowl" as a real fact or object. However, let's examine it in a bit more detail. We realize that although we can argue that it is not true that this apple exists in reality, we must consider that the dream of this apple in the fruit bowl is real, and that, therefore, since the field of meaning of dreams exists, in this field we must accept that this dream is a reality.

Gabriel's broad-spectrum realism is a very inclusive philosophical doctrine that encompasses science and its explanations, as well as other forms of knowledge or appreciation of reality that differ from science, such as art or theology. Note that, by not speaking of a single reality, the apple in the fruit bowl is constituted as an object or fact by its location in countless fields of meaning. In line with what I have been thinking, I believe that it is not only possible to think from science, but that science does not represent the only way to apprehend reality, especially when it is presented under infinite fields of meaning.

3.19 Sensitivity to Reasons

It could be considered a prejudice to suggest that we should educate ourselves to be sensitive to reasons. However, this is not the case, as education in this area holds the same status as education in many other areas, such as respect for others, sensitivity to people's pain, or appreciation of beauty. Indeed, all these matters—and many others—should be included in the educational package from childhood. But let me break a lance in favor of sensitivity to reasons, because I consider it central and fundamental to achieving sensitivity in all those other areas where it is also desirable to be sensitive.

Being sensitive to reasons means that when we are faced with any issue—in a conversation with friends, with strangers, with professionals, in any forum, or when we follow as spectators any discussion or discourse between third parties—we are already prepared to discern and appreciate the coherence of the discourses presented to us. We cannot demand coherence and logic in the speeches of others without, at the same time, demanding it of ourselves when we construct them. I refer to "the internal logic and coherence of the discourse," assuming it exists. Because what is being presented to us may be a simple "I opt for this or that," "I like this or that better," or "That seems better to me than the other." In our day-to-day dealings with others, we are used to hearing phrases of this kind, which, at first glance, seem to confer autonomy and prestige on those who formulate them, but which, in reality, reveal little or no sensitivity to reasons.

The autonomy of individuals to formulate an "I like or dislike," "I opt for this or that," "for this or that political party," etc., is a wonderful and necessary condition in free societies, where the individual can exercise their right to make a "final" decision on any of the matters mentioned above. But note that I have indicated a "final" decision. This cannot be a simple appetence—a process without internal deliberation, without reflection or maturation, without criticism, in a word. Let us say that we must add one more condition, which would be sufficient for this free society to function properly: that decision-making—the simple "I choose this or that"—be preceded by reasons and reasoning, and not simply by emotions.

Should this attitude of sensitivity to the exercise of reason be present even in the smallest of daily affairs? Strange as it may seem, I believe it should be. The time we must spend on any matter before making a decision will not be the same; some will require only seconds, others hours, and others a considerable amount of time. However, the necessity of adhering to reasons precedes this approach to decision-making.

Now, we are not born with a sensitivity to reasons, nor with a love for the exercise of reason; we must educate ourselves for it. Moreover, we may be too lazy to string together arguments to reach a reasonable conclusion. Many other characteristics accompany human reason, some of which may contradict what reason dictates to us, or, more simply, we may not have the disposition or habit to think. We often find ourselves torn between the dictates of reason and those of passion—the heart. In this respect, I believe these oppositions are mostly fallacious: passion is blind. It does not follow any reason; most of the time, passionate decisions lack reasoned foundations. It is much more interesting to combine reason and passion—in that order.

And now, let's go back to the beginning, because what better tool to cultivate sensitivity to respect for others, to the pain of others, or to beauty, than to learn and develop sensitivity in the exercise of reason? Reason, ethics, and aesthetics are more closely linked than they might seem at first. Education in values (ethics) and beauty (aesthetics) can be achieved more effectively through a solid education in sensitivity to reason.

3.20 The Spontaneous Philosophy of the Scientist

Reflecting on the need for scientists to go beyond their professional activities and on the importance of both private and general education, a text by Louis Althusser came to mind. In 1967, the philosopher gave a course at the *École Normale Supérieure* in Paris entitled "Philosophy Course for Scientists." With some additions, the content of this course was published in French in 1974 and a year later in Spanish (Althusser 1974).

The first thing that struck me when I reread this text was its capacity to attract scientists at the time. Althusser noted the attendance of scientists from various disciplines, including physicists, chemists, mathematicians, biologists, and others from the human sciences. The initiative for the course came from him, from a philosopher, as he himself states. And why this course? Althusser justifies it: "Well, by working on the texts of the history of philosophy and the sciences and dealing frequently with our scientific friends, we have formed some ideas about the relations that philosophy necessarily maintains with the sciences. Or, better yet, a certain idea of the relations that philosophy should maintain with the sciences to be useful to them instead of dominating them. And even more: because we have formed, within the framework of an experience external to philosophy and the sciences, but indispensable to understand their connection, a certain idea of philosophy capable of being useful to the sciences."

I find it fascinating that Althusser concludes he can offer an idea of philosophy capable of being helpful to the sciences—and indeed, he succeeds. Let us examine the reasons. He organizes the text into three courses and an appendix. In the first course, he develops the notion that philosophy has no object—no specific object—as the sciences do. Philosophy does not produce knowledge but formulates theses. In fact, throughout the first course and much of the second, he goes so far as to formulate twenty-five theses for the general understanding by scientists of what philosophy does. The twenty-fifth is the thesis of the spontaneous philosophy of scientists. He writes: "In their scientific practice, the specialists of the various disciplines 'spontaneously' recognize the existence of philosophy and the privileged relationship that philosophy maintains with the sciences. This recognition is generally unconscious but can become partially conscious in certain circumstances. Although this second case appears wrapped in the typical modalities of unconscious recognition, these modalities constitute the spontaneous philosophy of scientists (SPS)."

Having made them explicit, Althusser develops the thesis in course three that scientists, whether they want to or not, have a spontaneous philosophy. In the philosopher's opinion, the scientist must know what this philosophy consists of, because it can be a philosophy "by default"—that is, it exists without being known. What better way to learn about our SPS than by attending this course or any other formative activity?

The appendix—actually a fourth course—is dedicated by Althusser to identifying the SPS of the prestigious biologist Jacques Monod, which he analyzes based on

Monod's inaugural lecture at the Collège de France. On many occasions, I have echoed a phrase of Monod's, which appears in his book *Chance and Necessity* (Monod 1971), and which I have engraved in my heart: "We must avoid the confusion of the ideas suggested by science and science itself." If we think about it, Monod is aware that science is one thing, and the ideas it suggests or underlies are quite another. He says that "we must avoid confusion," which is not the same as saying that "we must avoid thinking about the underlying or emerging ideas of science," because, in reality, these ideas always exist. It is worth dusting them off. Courses such as Althusser's are important for the formation of a scientist.

3.21 Science and the Liar's Paradox

What are we? Why are we here? These are not questions from which we can easily escape. I argue for the importance of incorporating scientific contributions into the discussion of these and similar profound existential questions that confront us. That arena has generally been reserved for the humanities, religion, and art. Well, science also has—I would say it should have—something to contribute in this respect. I should note, of course, that these are contributions from professional practice, because reflection on such matters is of common concern to all of us: some of us occasionally, probably at critical moments, and others—the most melancholic—permanently.

The arena of existential discussion is a field to which we must all come equipped with our corresponding intellectual tools, especially those of reason, and where, of course, no one can arrogate to themselves the right of exclusion or veto, or pretend, *a priori*, to possess the definitive or ultimate truth about what is being discussed.

Bringing science into the discussion of existential questions is what essayist Curtis White calls for in his book *The Science Delusion: Asking the Big Questions in a Culture of Easy Answers* (2014). One such question is the debate about the meaning we can give to the human species. From science, we could affirm, as of today, that we are a lucky product of biological evolution—that just as we are here, we could not be here—and that, in any case, there are no forces or laws that govern us (this term should not be taken too rigidly) other than those that science—physics, chemistry, or biology—has been discovering over the last four hundred years. However, we can also affirm from science that we are singular entities, capable of having found out, like no other, how it is that we have arrived here.

But this double condition regarding our status—namely, that we have nothing particular about us, yet at the same time we are genuinely unique—leads us to a paradox, which White formulates as follows: "I am special because I know I am not special." This is yet another variant of the liar's paradox.

In its simplest form, the liar's paradox can be formulated as follows: "This statement is false." If we assume that the statement is true, then what it says is true. But since the statement asserts that it is false, it must be false. Therefore, if we suppose it is true, we reach a contradiction. Suppose the sentence is false; then what it asserts must be false. Since it claims that the sentence is false, it must be true. Then, if we suppose it is false, we reach a contradiction again. Examine this other sentence and arrive at the double contradiction: "A person claims he is lying. Is what he says true or false?"

Upon evaluating White's formulation, we will soon appreciate, in effect, that we are faced with a paradox. Let us consider that the sentence is accurate and that I am special. But the sentence states that I am not, therefore, special. We enter into a contradiction. On the other hand, let us now consider that the sentence is false—that what it asserts is untrue—and that I know I am not special. But since what it states is false, the sentence is true: that I am special. We enter again into contradiction. Hence the paradox.

This is in no way a play on words. Science does not appear to be in a position to resolve this paradox at present. However, some of its most fervent disseminators want us to understand that there is no such paradox, simply affirming from science that there is nothing special about us. But let us return to the arena of discussion of the great existential questions. Others will come forward to offer their reasoned visions about the place of humans in the cosmos—with philosophical, theological, or artistic theses that are well-founded—and that will most likely go in the opposite direction, namely: we are special.

3.22 Scientism, Dissonance, and Distortion

It is not the first time I have tried to find points of convergence between science and art. The most relevant, I have no doubt, is the process of generating knowledge or discovery, which is always unfinished. It is not only that the world to be explored is infinite, but that what has already been explored and explained is now interpreted in the light of a new theory.

Some issues traditionally outside the scope of scientific explanation become explained by it, generating a "dissonance" with how they were previously understood or perceived. But also, within science itself, we move from one theory to another—dissonance again—and what was once interpreted in one way is now interpreted in another, or the latter, in some way, engulfs and expands the former. Dissonance is an excellent term to characterize the dynamics of science, although it should not be forgotten that these dynamics are always framed within an open-ended path.

Art also admits to this process of consecutive dissonances in its history, and in response to certain past currents, new ones emerge that reinterpret the artistic object. And this process does not seem to have an end—nor should it be able to have one. From the perspective of dissonance and open paths, science and art intersect. They can understand each other and even nurture each other.

Let us now consider what we are facing if, in both science and art, we are presented with closed roads or roads that can no longer be traveled. What or who determines that these roads are no longer viable? We need not think of impositions external to science or art. There may be internal agents that encourage this closure. One is the belief that science is the only means of accessing knowledge of reality. Science harbors within itself its worst enemy: scientism, the ideology of science itself.

Under the umbrella of such an ideology, it will not be possible for us to communicate with artists or to have discussion forums where we can share our findings. If we adopt an open position on how to approach the study of reality, we may accept that there is a long way to go before we can find explanations of the world that are still far off—and that others are exploring with different kinds of tools.

For example, considering ourselves a machine may be an excellent metaphor. But are we a machine? We cannot say, at this time, that we have such a demonstration. It can be a metaphor as powerful as it is dangerous. We can trace this metaphor of living beings as machines far back in the history of science. However, we should not be confused, because the gap between metaphor and reality is still vast, and if we do not appreciate it as such, we are faced with a distortion.

Artists also interpret or recreate reality; we can find common ground if we admit that we also work with metaphors. For the artist, navigating metaphors within the creative process is second nature. And, of course, if any official current of thought is imposed on their creative process, the artist will no longer feel like an artist; they will work against it to continue considering themselves an artist.

In different ways, as can be seen, the distortions that operate in the paths of science and art can lead to—or presuppose—ideologies that should remain outside them, because these distortions hinder the proper dynamism and openness that characterize both. Scientism is the ideology of science that considers that only science has the true answers to explain things, often proposing to society mere ideas, assumptions, or interpretations that are not supported by science itself.

But it is not as simple as it may seem for the scientist to move away from scientism—an ideology that sweeps and erases. It sweeps, removing as if they were debris the other forms of thought by which we interpret, explain, or live in the world. And it erases—or tries to erase—that debris until it disappears, because it does not consider that it contributes to the proper knowledge of things. Scientism favors a single way of thinking.

3.23 Humanism and Science

If we start at the end, we will agree that science is an institution that serves the productive apparatus of nations and the private sector, from small companies to multinationals. With all the evidence in favor, we can affirm that science is the most powerful ally for economic growth and progress. However, I will not go into what economic growth and progress mean. For one reason, these two terms are insufficient if we do not specify who the beneficiary is. In other words, we must analyze them and determine to whom they refer: the nations or the peoples? The citizens?

The whole of humanity? The debate is served if we were to enter, as we can, into measuring whether all these collectives benefit.

With some bitterness, although open to debate, I maintain that science is probably the social institution that provides the greatest welfare to all humanity—that is, to every member of our species. In a supposed race to see who wins in this laudable endeavor, I affirm that science, compared to other institutions, wins by a distance. Note that I am formulating this in relative terms. I am not claiming that science brings absolute welfare. Throughout history, various groups have faced the negative consequences of science. But, I insist, from the perspective of increasing perfectibility—that everything can be improved—science leads the contribution to human welfare.

But the science I have spoken of here is positive science, when, in reality, science is much more, because it should not only be examined from the perspective of its contribution (or lack thereof) to human welfare or the safeguarding of the planet. It is essential to highlight this point because it is very common to criticize science from an exclusive perspective that assumes it serves the interests of a few rather than the entire community. When science is equated with economic growth or progress, we risk overlooking what is genuinely scientific: an endless openness to the world's knowledge.

It is necessary to remember again and again how it arose and what is behind that movement that, during the sixteenth century, forever transformed our way of understanding and explaining the world. The positive in science is an addition that has run parallel to its development and has become so crucial that no self-respecting public or private institution fails to resort to it. But there is a more profound, more fundamental sense—peculiar to each individual—of what it means to do and know science, which I think is feasible to consider under the umbrella of humanism. Science as a means of liberation through knowledge is antithetical to the idea that science and humanism are incompatible.

Rather than talking about a knowledge-based economy or the progress of humankind due to science—or, in other words, positive science—we have science as it is, the one that helps us or makes us question any issue. In that sense, it is critical even of positive science itself. It is not its aim, even if it succeeds to a greater or lesser degree, to promote well-being. Therefore, educating ourselves in science—not necessarily in positive science—opens a path to conciliation with humanism, including the most belligerent and critical one that always positions itself at the service of the various powers.

In many places on the planet, the most detached and kind individuals we can imagine are doing science with no other interest than to pursue explanation or discovery. Additionally, in any location, no matter how remote, many others are trying to understand and explain what scientists are after. Under this double perspective—that of the one who does it by vocation and that of the one who follows it by devotion—science instruction becomes necessary. Liberation through science is the responsibility of both, and both must be educated, not because we are going to be elements that contribute to the chain of production, consumption, growth, or progress, but rather because science—without a predominant position over other forms of knowledge—offers a unique path to understanding. It does not conform to reality to portray science as an institution exclusively devoted to the powers that finance it.

3.24 Genealogy of Thought

I am reading "*The Meaning of Thought*" by the German philosopher Markus Gabriel (2020). I still have a long way to go to finish it—not so much because I have only read a few pages, which is not the case—but because each page invites reflection. I stop to think about the ideas the author offers, which help me reconsider or refine my own, especially those related to science.

The fact that Gabriel's work is about thought and its meaning, and that I try to reflect on thought in the light of science, leads me to believe that both approaches share a commonality. That common ground is the human being himself, because thought and science are two genuine products of human activity. From my perspective of "thinking from science," I can easily understand the genealogy of human thought. For Gabriel, on the other hand, thought is something genuinely human—a kind of sense, although increasingly exposed in recent times to its destruction through disuse. To stop thinking, to suspend thought, is more straightforward than it may seem, and doing so, in fact, leads inevitably to the extinction of humanity in favor of other kinds of entities. But, being conscious entities ourselves, how can we abandon our most outstanding quality?

Gabriel formulates two laws, which he calls anthropological, that help us understand the threatening danger to which I am referring. In his first law, he maintains: "The human being is the animal that rejects its condition as such." I would reformulate the law as follows: "The human being is the only animal that can reject its condition as such." Assuming our animality—that it has become part of our nature—is one of the significant contributions of science. Previous theses have always positioned the human being as an entity free of animality. But science also unravels how we came to develop the thinking that characterizes us. We can trace the genealogy of the evolutionary background of animal cognition. I would say: the winding path through which thought has reached the human species.

On the other hand, rejecting our animality is a dubious aspiration. I would not know whether such rejection is an unspoken desire on the part of some to return to pre-scientific times—simply despising it—or rather, on the part of others, to aspire to the same, but on a scientific basis that is not fully established. Both positions, as I say, may wish to dispense with it, to eradicate it, but it is not possible to eliminate, just like that, our real animality. In other words, any attempt to destroy our animal condition—which is nothing other than our biology—requires a knowledge that, in my opinion, is far from complete. Heidegger (1996) already characterized well how science, in this case science about ourselves, proceeds: by successive and multifaceted approximations, never starting from a fundamental apprehension or intuition of what could be.

It may help to understand my argument to explain Gabriel's second anthropological law: "The human being is a free spiritual being." This thesis could give the impression that it transcends science. Still, I have doubts, because both the concept of spirit and that of freedom admit a rational interpretation by the author. For Gabriel, the spirit represents "the capacity to live life in the light of an image that reflects what the human being is." And to be able to realize it, to substantiate it, constitutes the dimension of freedom. Who can represent themselves and also be free, other than a human being?

Now let us consider where we might end up if we dispense with thought—or, much worse, if we go beyond our biology when we do not yet understand it—and recreate ourselves as beings that are no longer us. In the hands of the second law—or, better said, in its absence—we are heading toward entities that probably cannot develop an image of themselves and, therefore, cannot be free. Acknowledging these questions does not weaken scientific inquiry, but situates it within a broader human effort to understand meaning and existence.

References

Althusser L (1974) Philosophy and the spontaneous philosophy of the scientists and other essays. Verso, London

Anders G (2014) Acerca de la libertad. “Una interpretación del a posteriori” seguido de “Patología de la libertad. Ensayo sobre la no identificación”. Pre-textos, Valencia

Brockman J (1996) The third culture: beyond the scientific revolution. Touchstone, New York

Burtt EA (1954) The metaphysical foundations of modern science. Doubleday and Company, New York

Castrodeza C (2009) La darwinización del mundo. Herder, Barcelona

Castrodeza C (2013) El flujo de la historia y el sentido de la vida: la retórica irresistible de la selección natural. Herder, Barcelona

Gabriel M (2017) Why the world does not exist. Polity Press, New York

Gabriel M (2020) The meaning of thought. Polity Press, Malden, MA

Godfrey-Smith P (2016) Other minds: the octopus, the sea, and the deep origins of consciousness. Farrar, Straus and Giroux, New York

Gould SJ (1981) The mismeasure of man. W. W. Norton and Company, New York

Haldane JBS (1927) Possible worlds. Phoenix Library, London

Hawking S, Mlodinow L (2010) The grand design. Bantam Books, New York

Heidegger M (1996) Einleitung in die philosophie. Vittorio Klostermann GmbH, Frankfurt am Main

Koyré A (1994) Pensar la ciencia. Paidós, Barcelona

Martínez G (2012) Borges and mathematics. Purdue University Press, Lafayette, IN

Monod J (1971) Chance and necessity. Random House, New York

Moya A (2010) Pensar desde la ciencia. Editorial Trotta, Madrid

Moya A (2018) La ciencia y el arte desde la perspectiva de la transevolución. Arbor 194(790):a477

Ortega y Gasset J (1982) Meditación de la técnica y otros ensayos sobre ciencia y filosofía. Alianza Editorial, Madrid

Pascual S (2017) Per qué filosofia? Neopatria, Alzira

Shirrmacher F (2014) Ego. Las trampas del juego capitalista. Ariel, Barcelona

Snow CP (1959) The two cultures and a second look. Cambridge University Press, New York

Trias E (1999) La razón fronteriza. Destino, Barcelona

White C (2014) The science delusion: asking the big questions in a culture of easy answers. Melville House, New York

Chapter 4
Science and Politics

Abstract In the next group of essays, I explore the complex relationship between science and political structures, particularly in Ibero-America. I argue that science education should be recognized as a fundamental right of citizenship, essential for fostering critical thinking and participatory democracy. Drawing on personal experiences and historical examples, I reflect on the lack of scientific training among political leaders and the consequences of underfunding research. I advocate for a more inclusive and equitable science policy that values both public and private contributions while ensuring universal access. I also examine the role of science in addressing global challenges such as climate change, digitalization, and technological dependence. Through these reflections, I call for a renewed pact between science and society—one that empowers citizens, respects cultural diversity, and ensures that scientific progress serves the common good. Science, I argue, must be governed wisely and shared broadly to help us navigate the future. The arguments presented here are offered as normative proposals open to discussion, rather than as definitive political prescriptions.

Keywords Climate change · Common good · Digitalization · Ibero-America · Participatory democracy · Science education · Science policy · Technological dependence

4.1 Scientific Training of the Leading Political Class in Ibero-America

I completed my postdoctoral work in the second half of the 1980s in the United States of America (USA) under the guidance of the well-known geneticist and evolutionary biologist Francisco J. Ayala. Many scientists have passed through his laboratory and have gone on to develop scientific research in various Ibero-American countries. For me and many others, there was a clear before and after in our careers following our time in his laboratory (Bermejo 2019).

A. Moya, *Reflections on Science and Evolution*,
https://doi.org/10.1007/978-3-032-25682-9_4

Now, however, allow me to reflect on an observation Professor Ayala made about the relevance of science as an engine of economic development. He told me that after World War II, the United States, driven by committed politicians, decided to support science of all kinds, substantially financing projects in the country's research and higher education institutions. The following conviction was part of their ideology: hand in hand with science, there will be greater national wealth, economic growth, and improved quality of life for citizens. He showed me a graph illustrating the near-perfect correlation between increasing public investment in science and economic development from the late 1940s to the 1980s.

Professor Ayala applied for competitive funding from the US Department of Energy to work with the fly *Drosophila melanogaster*, examining how different mutants with varying abilities to transform oxygen radicals responded to specific doses of radiation. Why would an entire Department of Energy be interested in funding something seemingly unrelated to energy production or optimization? He held that scholarship for several decades.

What lesson can we draw from this regarding the public funding of science in Ibero-America? Is it as simple as demanding more scientific investment from the political establishment? Unfortunately, no. The political establishment in my country—and in Ibero-American countries in general—is far from believing in the beneficial effects of science on the economy and welfare. Note that I am referring to the political ruling class. Who are they? I am not referring to the advisors or committees that assist the leaders. Our administrations are full of High Advisory Councils and Academies that could help the political ruling class make informed decisions, as is the case in the most scientifically advanced countries. However, in those countries, the political class tends to be more receptive to science. This belief stems from a secular tradition that has shown how ideas or theories aimed at understanding and explaining the nature of things—even without immediate application—have led to products and technologies that generate wealth.

There is a peculiarity in the political class of countries that have advanced in science that I do not see in ours. Our politicians have their own priorities, which certainly do not involve "believing" in science, because it has not served as an engine for economic development in the recent history of their respective countries.

Why don't we do a little research? How many of our presidents, prime ministers, governors, ministers, or secretaries of state have a scientific education—such as a doctorate in physics, chemistry, mathematics, biology, or any branch of engineering? I hypothesize that the number is relatively low. I suggest that someone study and compare the challenges facing science in Ibero-America with those in advanced countries, taking into account the scientific training of their respective political ruling classes. I can think of no better way to reverse the underfunding of science than to have a political ruling class educated in the hard sciences.

4.2 The Galileo Schism

In 1965, Friedrich Dessauer's study on the Church's trial of Galileo was published in Spanish under the title *El caso Galileo y nosotros*. The work had initially appeared in Germany in 1942, amid World War II. Dessauer was a quintessential, enterprising scientist who had to emigrate from his country during the Nazi era and who devoted himself, like many other German intellectuals of his time who also emigrated, to a wide range of university teaching and lecturing activities.

Dessauer, a Christian, was fascinated by the Galileo case because of how the Church opened a rift with science through the trial of the Italian physicist. This rift has had far-reaching consequences in many areas of Western history. Dessauer situates the Church's process and resolution against a faithful believer like Galileo as being motivated, probably, by fear of new splits within its ranks. We are referring to a period when the Thirty Years' War had solidified the division between countries that had adopted the Reformation of Luther and Calvin. It was precisely the movement initiated by specific individuals—whose apparent intention was to return to the Church's authentic values—that ignited not only the fuse of the separation from Rome but also the configuration of many modern European states. Dessauer argues that the Church did not want to face the consolidation of a new schism.

It is surprising, however, that of the three major schisms in the Church—namely the Orthodox Church, the Reformation, and the hypothetical Galilean schism—the first two resulted in the creation of Churches separate from the Catholic Church, but they remained Churches nonetheless. Now, what was the effect of Galileo's supposed schism? In my opinion, it has had profound consequences. Galileo inaugurated modern science as a way to approach knowledge and explain the world. Catholic thinkers and theologians, both before and contemporary to Galileo, would not have understood—or would have misunderstood—the kind of knowledge of the world that Galileo pursued with total conviction. He was a passionate man of formidable intelligence and unwavering convictions.

Galileo was closely connected and had strong friendships with the Papacy and with the heads of several academies and universities of the time, where the Aristotelian vision of the universe prevailed. This vision, which the Church had adopted as a rational framework to support faith in the existence of a God who created an ordered and perfect world, was also prevalent. Galileo managed to convince his learned peers, and probably nothing special would have happened if natural philosophy had admitted science—as it was beginning to emerge—as a valid way of knowing the world.

We are several centuries removed from that case, yet we continue to grapple with the same fundamental questions. Some try to solve them through faith, and others do not. But the truth is that science, in its strictest sense, also guides and supports us in answering these fundamental questions. The scholastic stubbornness of the Church—which was not limited to the Galileo case but persisted afterward—promoted a significant schism of a very different nature from the Orthodox and Protestant ones: the schism of science.

The Catholic Church, which had been open-minded about Renaissance and Baroque art and culture—perhaps to promote its own glory and power—made an exception with science, with Galileo as one of its representatives. Natural philosophy was a scholastic discipline, and if we consider the intellectual disposition of some of its most prominent practitioners before, during, and after Galileo, we can conclude that they were open to incorporating science as a way of understanding, explaining, and harnessing nature in an effort to promote a better earthly life. But they were ultimately overcome by others who believed they saw in Galileo's science a fundamental danger: that it would shake the natural order of the world as established by Aristotelian scholasticism.

Galileo's inquisitors feared that his ideas would be schismatic, or that allowing them to develop within the Church would create a new separation among his followers. That fuse was lit, with the well-known effects it has had since. Before and during Galileo's time, Italy was the cradle of the greatest artistic and scientific innovations. It then took a back seat—at least in the realm of science. The countries that embraced the Reformation were not so critical of science; quite the opposite, they promoted it. This is one of Dessauer's most relevant theses: the divergent effects on the development of science in countries where the Catholic Church did or did not hold social and intellectual power.

Nevertheless, I dare to offer a final reflection. Strictly speaking, if we consider the ultimate goal of science, we are compelled to engage in dialogue with the Churches, because the fundamental questions that concern us remain the same—before, during, and after Galileo.

4.3 Science and Ideology: On the Subject of Intelligence

In 1981, the renowned paleontologist, scholar, and science popularizer Stephen Jay Gould published his work, The Mismeasure of Man (Gould 1981), whose Spanish version appeared a few years later under the attractive title *La falsa medida del hombre*. At the bottom of the cover of the Spanish edition is a synthetic phrase—almost certainly added by the publisher—to encourage readers to engage with the book. The sentence reads: "A devastating attack on the biological determinism of intelligence." If we combine the book's title with this summary sentence, one might think we are dealing with a work that irrefutably demonstrates that intelligence is not genetically determined. In reality, this is not the case. Gould explains the uses and misuses of a supposed science practiced by certain scientists interested in the complex problem of intelligence.

If we are so drawn to the issue of intelligence, it is because we believe it defines us. It is our uniqueness that differentiates us from other species. We are different—but the same is true of any other species. They all have their own uniqueness. Throughout history, we have encountered multiple definitions of what constitutes humanity. Without pretending to be exhaustive, it has been said that we are an "imputative biped," "a political animal," "a being that laughs," "a being that thinks," "that judges," "that works," "that creates," to cite just a few. But none of these definitions can escape what we might call "the omnipresent biological component of our uniqueness" or simply "our nature." It is also true, if we follow this reasoning, that if our uniqueness has a biological component, it must also have at least one other—a cultural, social, or environmental component, or a combination of all three. This leads us to conclude that to understand what defines our uniqueness, we need to measure both biological and sociocultural aspects. Whether the biological part determines us and the sociocultural part does not is a far-reaching question that still requires extensive research.

So, what does Gould intend with his study of intelligence? The definitions of man I mentioned above are not scientific formulations. Nevertheless, they attempt to capture, in a few words, what we are and the value of conciseness, often formulated by eminent philosophers. However, Gould directs his meticulous gaze toward certain purportedly scientific studies on intelligence carried out throughout the nineteenth century and the first half of the twentieth century. These studies were conducted by influential scientists whose conclusions were used, to a large extent, by the political ruling class to justify discriminatory measures against social and ethnic groups in the United States and some European countries. Depending on the case, Gould demonstrates that there is little actual science in these studies, that the applied science is flawed, or that there is both manifest and unintentional falsification or manipulation of data. What underlies these studies, in short, is a range of attitudes among certain evolutionary biologists, psychologists, and statisticians—from prejudice to ideology—that they used, under the guise of scientific objectivity, to justify what can generically be called "racism based on intelligence."

When measuring something as intangible yet real as human intelligence, we run the risk of drawing false conclusions if we are not adequately cautious. Scientists are as human as anyone else, and preconceptions, ideologies, and prejudices affect us too. For certain issues, the effects of such predispositions on scientific work can be very significant, to say the least. However, Gould does not directly address the question of biological determinism of intelligence. He reveals the errors and ideological biases of some who tackled the problem. He eloquently describes his work as an act of demystification of positive science. But the problem remains: Who said science is over?

4.4 Which University Is in Crisis?

I am a professor of genetics at the Universitat de València, an esteemed academic institution with origins dating back over five hundred years. Richard Dawkins, a distinguished and well-known popularizer of biological evolution, was awarded the title of Doctor Honoris Causa by my university in 2009, during the Darwin Year. Prof. Dawkins was impressed by the ceremonial grandeur and attention to detail of the naming ceremony. I acted as his godfather—a figure included in the ceremonial protocol—and, among other duties, I had to accompany him. At the same time, we waited to be called by the faculty to begin the appointment ceremony in the university's magnificent and historic auditorium. During this wait, Prof. Dawkins remarked on the resemblance he observed between his university, Oxford, and mine. His surprise stemmed from being familiar with the pomp of British academic ceremonies, which he attributed to Oxford's antiquity. He had not imagined that my university could be similar. My response, which also surprised me by its immediacy, was: "Richard, my university is as old as yours; you should not be surprised by the resemblance."

Everything in history, including academic institutions, undergoes evolution. One can begin, for example, with ancient universities whose origins go back centuries and observe that, while they may have been similar in their initial objectives and equally effective in fulfilling them, they have subsequently experienced a series of circumstances that have altered—not so much their goals—but rather their ability to fulfill them adequately.

It has long been said that the university is in crisis. My perception of this issue is different: I believe that it is not the university itself that is in crisis, but rather certain universities. Despite the tremendous social and cultural transformations that have occurred since the first university institutions were established in Europe, the truth is that the supreme values pursued by the academy have remained largely unchanged. Only some universities—and there may be many—fail to achieve these goals. There are better and worse universities, and the worst of them have most likely lost sight of their foundational mission.

Let me recall these core values by reflecting on my own university. The Universitat de València is known as the *Estudio General*. What an outstanding designation. Although the name originates from certain educational powers granted to it by the city of Valencia, the institution's objective—as any self-respecting academic institution should have—was that of general study: undirected, broad, deep, multidisciplinary study, pursued with open-mindedness and grounded in creativity.

I am writing this from the city of Boston, where I have traveled to meet with colleagues working at the Massachusetts Institute of Technology (MIT) and Harvard University. There are various rankings used to evaluate universities, each employing a wide range of criteria to compile quality lists. These two universities are currently ranked first and second in the world, respectively, and those that follow, in my opinion, adhere to the same philosophy. When one thoroughly explores their departments, schools, faculties, and research institutes, one observes that they are institutions with a profound vocation for "General Study," where excellence, creativity, and intellectual rigor form the inalienable foundations that sustain them. These are the same founding principles that have always characterized the university. While it is true that the university has required rethinking throughout its history, such rethinking has ultimately consisted of a return to its origins. It is a different matter that many universities today retain only the name, having strayed so far from their principles that we can no longer recognize them as such. Of course, these so-called universities are found at the bottom of the rankings.

4.5 Vindication of Ibero-American Science

Historians, let us recover Ibero-American science and, with clear and illuminating insight, make visible to the world the achievements of scientists who have belonged to sectors as varied as civil and military administrations, the Church, universities, companies, as well as liberal professionals such as lawyers, doctors, pharmacists, and others. I admit to being persistent, but I must confront the specter. I do not know if I can affirm with more than probable certainty that we lack a scientific tradition, that there are no firm foundations to sustain it, that our science is barely established, that our contributions are relatively recent, and that they largely result from a learning process in countries with a long-standing scientific tradition. However, I prefer to think that this asymmetry is fueled by ignorance. That is why our historians must dust off the archives and return to the documents detailing everything that has been done since at least the colonial empire, which began and consolidated with a broad research and naturalistic endeavor associated with the Iberian Peninsula and the Americas. Our scientific history must be recovered so that we can properly insert ourselves into the fundamental concept of "tradition" in science. Let us consider a tradition that I wish to define as the Ibero-American scientific culture.

Spain, for example, has experienced dark periods in science which, conversely, may correspond—I venture as a hypothesis—with the development of science in what were once great overseas capitals, such as Lima or Mexico City, or the nations they represent after achieving independence, to cite just two among several possible examples. Let us adopt this perspective to avoid sinking into self-inflicted misery, which would not align, if accepted, with the significant Ibero-American scientific developments that have occurred since the fifteenth century.

And so, how many contributions should be restored? How many unrecognized discoveries should be rightfully reclaimed? How important is science to be recovered and vindicated? A great deal, I fear. I am unsure whether the enterprise will succeed or prove worthwhile. Still, it would likely be far worse to abandon it, thereby perpetuating the unfair asymmetry of the supposed non-existence of scientific tradition in certain cultures.

It is essential to rediscover the history of Ibero-American science and bring it to light, far beyond isolated centers, institutions, or the actions of individuals with limited visibility. It must be reclaimed and made public internationally. And the same dimension I claim for Ibero-American science, I would also like to extend to any other cultural community, since that is where the true universality of science lies. I doubt—however slightly—that there exists any culture without a scientific tradition. Although there are degrees and levels of excellence in achievements, science emerged as a rational practice with a method only a few centuries ago. Its development has been gradual or accelerated, depending on the culture. That is why we must always vindicate it—to recreate tradition and avoid the feeling, often fed by others, that certain cultures are not inclined to promote science.

And let us bring to the international stage, I repeat, those achievements that constitute our tradition—or the tradition of other cultures—regardless of whether there have been more or less adverse periods. This is how we can avoid being overshadowed by the brilliant and glorious traditions of countries that are currently very advanced in science. In no way do I undervalue the tradition of countries and cultures that founded successful academies of science or created magnificent institutions of higher education over the last three hundred years. However, if we dedicate ourselves exclusively to considering or studying those traditions—which they have learned and passed down well—we may reach the somewhat misleading conclusion that those are the only traditions that support modern science.

4.6 Science for Participatory Democracy

The Athenians met in the agora to discuss their laws and decide the political future of their city. Those who mastered the art of oratory and persuasion could participate and win the debates. It is not a reprehensible art as long as its practice is based on the exercise of reason aimed at the pursuit of truth and the common good. Otherwise, oratory becomes the art of demagogy. In the Athenian agora, there also appeared—mainly as a reaction—those who believed that reason and wisdom should prevail over demagogy. Oratory in the service of reason was valid, but not reason in the service of oratory—persuasion detached from truth or the common good.

In today's society, thanks to communication technologies, we can envision a global agora where we participate in decision-making. But how can we protect ourselves against demagogic oratory? And, perhaps more importantly, how can we participate in decision-making by appealing to reason in the pursuit of truth and the common good? New technologies fuel the transition from delegated to participatory democracy, but at the same time, they also expose us to demagoguery.

I maintain that science education is fundamental to shielding us from demagogy and making us free citizens—active participants in critical decision-making within a participatory democracy. But how much science education is needed to achieve these objectives? We will have to apply criteria to select, from among all possible areas of instruction—likely unmanageable in their entirety—that part which enables us to have a critical presence in a participatory society. As Heidegger (1996) states, with the birth of science, a new and inalienable way of approaching and engaging with the world emerged—distinct from and complementary to previous non-scientific modes, such as philosophical, artistic, or theological ones.

The need for science education in today's world is radical because it is thanks to science that we sustain ourselves and survive. A lack of understanding of the entities and concepts science provides is equivalent to a lack of the critical capacity necessary for informed decision-making. Other ways of knowing the world are essential and offer varying degrees of existential relief. But science's understanding of the world provides guidelines for action in an environment increasingly shaped by its products, including technological ones.

The corollary is that both scientific and non-scientific modes are necessary. Yet, while philosophical, artistic, and theological modes are totalizing, the scientific mode is cumulative and detail-oriented. Changes in intuition, reconstructions, and fundamental worldviews are the norm in the former; in science, the norm is the understanding and explanation of details.

Science is a more universal communal currency in a participatory democratic society than non-scientific modes. It is more an element of exchange, discussion, and consensus, arising mainly from outside influences on the individual. Non-scientific modes emerge more from the inside out, are more intimate, and contribute to shaping the radical singularity of each person. For both its singularity and commonality, the citizen must equip and educate themselves in both modes to be an active and critical participant in a democratic society.

4.7 One Thing at a Time, Unfortunately

The first steam-powered ship sailed in Barcelona on June 17, 1543—two hundred and sixty years before the American Robert Fulton developed and tested his prototype on the Seine River (Paris) on August 9, 1803, becoming famous as the inventor of the steamboat. The naval officer Blasco de Garay convinced Emperor Charles of the value of his discovery, and a commission chaired by Alonso de Rávago, treasurer of the Kingdom, was appointed to study the proposal's feasibility. Garay retrofitted a merchant ship with sails and oars, the *Trinidad*, equipping it with a device that generated steam and produced enough pressure to move the vessel through an intricate combination of internal wheels and external paddles.

Rávago did not recommend the invention to the sovereign, citing numerous problems: low speed, especially when the ship was loaded, frequent need for repairs, and the serious risk of boiler explosions. Spain was at the height of its imperial power—a country well-positioned to implement new ideas and projects and, ultimately, to advance science. However, it seems that the technology of the time was not sufficiently developed to overcome the problems Rávago rightly identified.

The mechanics and metallurgy required to build a ship of this kind—using, for example, the steam technology later developed by James Watt—help explain the time gap between Garay's invention and Fulton's. I often ask myself: Can you imagine what maritime technologies we would have today if steamships had begun development in the 1540s instead of the 1800s?

In a similar vein—though the underlying story is quite different—we can consider the rediscovery of genetics. The pioneer of genetics was the Augustinian monk Gregor Mendel, who conducted the first experiments on the inheritance of traits in his abbey in Brno. Mendel formulated a series of laws and hypothesized the existence of units of inheritance, later called "genes," whose existence would be confirmed a few decades later. But his work, published in 1866, was neither read nor understood by the scientists working in the leading scientific centers of his time. Consider, for example, that Darwin was Mendel's contemporary. Darwin's theory resonated throughout the scientific world when he published "The Origin of Species." Mendel's work, however, was rediscovered independently by several scientists in 1900, thirty-four years later. Can you imagine how genetics and molecular biology might look today if research on genes and DNA had continued uninterrupted from Mendel's time?

Ideas and theories—whether empirically demonstrated or not—as well as technical inventions and prototypes, may exist in the imagination of many individuals throughout history. Still, they only become substantiated when the right circumstances arise. Indeed, one thing at a time, one might say. But there is a moral—several, in fact—in the two examples I've cited regarding the progress of science. Scientific advances become evident when the theoretical and technical conditions are in place for them to be substantiated and realized; in short, when the "grapes are ripe."

However, we also observe—not without some pain—that certain advances should have occurred at a particular moment (the grapes were ripe), but did not, due to external factors, including sociological and cultural ones. And worse still: how much science and intelligence have we lost throughout history because those who could have advanced it lacked the conditions to do so? The delay of science is a tragedy for humanity.

4.8 Backwardness in Science Is a Disgrace to Humanity

Many other activities of our species are as important—or even more important—than science in promoting development, welfare, or progress. However, I must admit that I assign it particular value, as it has likely contributed more than any other to these objectives. Several years ago, I had the opportunity to travel to different cities in Mexico to give lectures at universities, research centers, and educational institutions. This activity was part of the commitment I had made to the Science Advisory Council of the Presidency of the Republic of Mexico after being awarded the Mexico Science and Technology Award in 2015. Particularly moving for me was the conference I gave to high school students at a center located in a marginalized neighborhood on the outskirts of Mérida (Yucatán)—a neighborhood that, according to the center's management, concentrated all kinds of social problems: family instability, violence, lack of community roots; in short, conditions that make it very difficult for a boy or girl to receive a proper education.

It was also emotional because the content of my lecture was, to a large extent, autobiographical. I showed the young students that I, too, came from a similar environment—that my family, immigrants with little or no formal education, had moved from the interior of Spain to the Mediterranean coast in search of a better life, and that they lived for many years—those corresponding to much of my childhood—in a marginal neighborhood with all kinds of problems, not unlike those the students might face, allowing for differences in time and place. I shared my story—I couldn't think of anything better—and explained how, with the help of my family, I was able to pursue my education, overcoming many obstacles, and eventually reach higher education, dedicating myself to science. I did not offer any recipe for escaping such circumstances, because I do not have one. I shared my story, hoping it might encourage them to pursue their education. What I did feel, however, was the enormous closeness between them and me at their age. That conference was a magical moment of inexplicable complicity and empathy.

But I must return to the issue of the lack of a prescription and, not without a certain pessimism, reflect on how many of those young people—like those who lived alongside me during my childhood and early youth—will ultimately drop out of school and miss out on the education that is rightfully theirs. It is not the fact of these individual educational losses that I want to address here—though they are an unmitigated tragedy—but rather how many of them, taken together, represent a significant loss for the progress of humanity.

Some recent texts suggest that human progress is unstoppable—that since our emergence as a species, we have continually advanced in parameters related to quality of life, wealth, education, and more (Pinker 2018; Ridley 2010). My reflection is not to dispute that progress, but to ask: What would that progress look like if all young people, past and present, had had access to education? In short, how much faster would we have advanced if, from the earliest times, young people everywhere had had access to the education that, in any time and place, only a privileged few have enjoyed?

How many brilliant scientists, artists, philosophers, and politicians have we lost along the way—not for lack of talent, but for lack of opportunity—whose ideas and actions could have accelerated humanity's progress far beyond what we observe today? This is not only a personal injustice, but an injustice against humanity. The extraordinary transformations that have occurred throughout history are ultimately the work of specific individuals—some visionaries, others executors of great ideas. But these individuals belong to the right tail of the talent distribution. The greater the total number of educated individuals, the greater the number of talented individuals who will emerge.

4.9 Science, a Universal Right

We know from the Universal Declaration of Human Rights that human beings have the right to various things, such as education, health, and justice, among others, which are well-known rights. But Article 27 of the Declaration also recognizes that: "Everyone has the right to participate in the cultural life of the community, to enjoy the arts and to share in scientific advancement and its benefits." It is not only that one has the right to participate in science, but also to receive its benefits. Throughout history, citizens have benefited from the findings of science. Science has a practical dimension, meaning that its results, sooner or later, lead to improvements in the quality of human life.

But the other dimension of the right to science is less obvious: the one related to participation in science, which, in my opinion, also has a dual aspect—instruction through fundamental science education, and ultimately, the possibility of engaging in scientific work or dedicating oneself to science.

The Universal Declaration serves as the basis for newer treaties that have expanded and detailed its articles, and which are binding on the states that ratify them. Thus, in one of the 1966 covenants—the so-called Covenant on Economic, Social and Cultural Rights—Article 15 recognizes "the right of everyone (…) to enjoy the benefits of scientific progress and its applications." This article also includes provisions for signatory states to take measures for the conservation, development, and dissemination of science, to respect the freedom essential for scientific research, and to encourage and develop international cooperation and relations in scientific matters.

These issues are the subject of an excellent article written for *El País* on November 18, 2018, by Mikel Mancisidor, Professor of International Human Rights Law at American University (Washington D.C.) and member of the UN Committee on Economic, Social and Cultural Rights. I bring this issue forward because the right to science was recently endorsed by the *XXVI Ibero-American Summit of Heads of State and Government* in the Guatemala Declaration, made public on November 16, 2018. In paragraph 45, the Ibero-American countries support Article 15 of the Covenant mentioned above and move toward the "effective recognition of the human right to science." Mancisidor titled his article *Science, an (old) human right with a future*, which is thought-provoking. Indeed, this right is already contained in the Universal Declaration—soon to celebrate its seventieth anniversary—and in other subsequent agreements that are more than fifty years old. Why has it taken so long for UNESCO or the United Nations, among others, to begin developing the discourse of science as a human right?

The Guatemala Declaration refers to working toward the "effective recognition of the human right to science." If it is stated that it must be "effective"—i.e., "with effect"—it is implicitly acknowledged that it currently lacks such effect, which is quite common among many rights in the Universal Declaration. Recognizing the existence of a significant, universal problem related to the lack of effectiveness of any right prompts the appropriate international bodies to advance its implementation and realization.

And if we apply this to the case of science, we can say that it is only now, having identified the gap, that we are beginning to implement it beyond foundational declarations. Why science, now, if it has always been there? In my opinion, we are facing serious global challenges that require our voices to be heard more than ever. And what better way than through the recognition—evident, yet long overdue—that science is a human right, and that states and governments should be compelled to take measures grounded in science to confront the pressing problems that afflict us now and in the immediate future? The human right to science has a great future, because without science, we are not guaranteed one. This claim is defended as a normative ideal grounded in democratic values, and not as a description of current political realities.

4.10 Shared Science in Ibero-America

I have had the opportunity to visit a good part of the Ibero-American countries—including Portugal and Spain—and to witness the vast richness of the collective culture they represent. Well, science is one more element to be added to that cultural cart, as Santiago Ramón y Cajal, our monumental scientist and father of neuroscience, once called it. The expression of Ramón y Cajal that has gone down in history is: "The cart of Spanish culture lacks the leg of science." But the exact phrase, which he pronounced in 1926 during the inauguration of his monument in Retiro Park in Madrid, is even more compelling: "Because the wheel of science is broken, the pompous chariot of Hispanic civilization has stumbled along the road of history."

This is an excellent metaphor, because a chariot can hardly move with stability if one of its four wheels is missing. And culture—Ibero-American culture—moves; it must move. The quality of a society is directly related to its cultural achievements and their preservation. If science is not one of the wheels of this cart, culture does not move at the appropriate speed.

However, allow me to discuss Ibero-American culture and its place within the realm of science. I am not going to claim that this culture is a pure entelechy and that what exists in Ibero-America is only particular cultures. On the other hand, I am not opposed to such a claim. But when one looks with perspective, one appreciates that the Ibero-American cultural conglomerate exists, and that its amalgamation is natural; it requires no special effort to observe the enormous number of common elements that bind it together.

Indeed, it is necessary to advance this program and achieve, as is promoted in Europe, a Europe of the peoples. Ibero-American nationalism is probably less resistant to the idea of an Ibero-American culture than European nationalism. Europe has had to advance more rapidly because it suffered two devastating world wars and has therefore striven to build—and continues to create—a culture that transcends and integrates the enormous diversity of its peoples.

It is surprising to observe the significant progress made in Ibero-America toward the creation of a unified, integrative, and cohesive culture. There must be a reason for this, even though recent historical circumstances have not compelled political leaders to promote Ibero-American culture as a unified whole. Yet it seems relatively more straightforward—or at least more feasible—to achieve this in Ibero-America than in Europe, despite the many parallels between them.

Let us return to science. If we already possess an amalgamated cultural treasure in Ibero-America, incorporating science into this cultural cart should be relatively simple. Science—and the language that conveys it—is even more universal than the other wheels of the cultural cart, which comprise other forms of knowledge. The promotion of science by the various Ibero-American states, as well as by Ibero-American supranational bodies, is something that should happen naturally and inevitably. The absence or slowing down of this promotional effort should be critically examined. Something in the cultural environment, including science, in Ibero-America is hindering adequate action in support of universal science.

I have always maintained that science will be universal when it is practiced by everyone, everywhere. This task is made easier when specific cultures evolve to become part of larger, integrative cultural frameworks. All in due time, of course, because it is the nations themselves—through their organization and inertia—that often prevent themselves from transcending and moving toward new domains, which are nonetheless natural and rightful: those of supranational cultures amalgamated with evident ease. Science plays a crucial role in this process.

4.11 Science and the New Politics of "The Terrestrial"

Practically in one sitting, I read with genuine delight *Dónde aterrizar. Cómo orientarse en política* by the French philosopher, anthropologist, and sociologist of science Bruno Latour (2019). I particularly appreciated that he describes it as an abrupt text, written in the wake of the U.S. withdrawal from the Paris Climate Agreement on June 1, 2017. Does Latour feel outraged by this event? One cannot simply apply denialism to climate change or isolate oneself from a problem in which one is an active contributor. Although the text is abrupt, as the author rightly notes, it is not written in anger, which would have turned it into a pamphlet. As expected from an intellectual, it is a thoughtful and considered work on the relationship between politics and the future of humanity.

In a way, President Trump's decision to withdraw the USA from the Paris Agreement prompts Latour to reflect—and I would almost say, with a certain liberating catharsis—first, on the crisis of contemporary politics in adequately addressing that future, and second, on the kind of new politics we need and the axes along which humanity must act—humanity as a whole—to, quite simply, survive with this planet.

I take advantage of this general reflection on Latour's book to focus more specifically on the role that science has played. I must continue to play a role in enabling the survival I just referred to. Curiously, Latour argues that science has a fundamental role in this new issue, which he calls the terrestrial. The new politics must be oriented toward the proper treatment of the planet, overcoming the antagonism between policies focused on safeguarding the local (my people, my land, my country, etc.) and those focused on the global (others, other lands, the entire planet). Our species' relationship with the world around us must go beyond the assumption that it is inexhaustible and unresponsive, as if it were an inert body subject to our exploitation. But the Earth does react—and it can do so in ways that may well exceed what is bearable for us.

What would we know about climate change without science? Science is our best ally in the fight against this problem. But Latour argues that science must shed certain ideological varnishes that have made it an object of criticism and rejection by sectors of society that are deeply committed—both politically and existentially—to defending the planet. In his view, the Galilean revolution was foundational for the emergence of science, but it also cast us into a limited conception of what truly "moves the Earth." The planet is more than a simple Galilean body—it is not just another object in the universe.

To a large extent, according to Latour, this limitation—the failure to incorporate other forces—is what prevented us, over the past two centuries, from foreseeing the changes brought about by human activity, from recognizing the finite nature of resources in economic models, and from developing technical models of planetary dynamics capable of predicting change over short timeframes.

As mentioned, Latour observes a growing disenchantment with science among many who are aware of the planet's severe deterioration. Rebuilding trust between this broad collective and science, he argues, requires the formulation of more systemic, multi-force conceptions capable of producing long-term planetary solutions.

I am not entirely clear on what Latour means by a "lack of rationality and objectivity in science" as a consequence of a purely Galilean conception of the planet. He argues that the forces of life must also be incorporated into our understanding of planetary dynamics. However, historically, biology took two centuries to mature and become a robust science, compared to physics. Only recently has biology acquired the conceptual and technical tools necessary to contribute to the modeling and prediction of planetary change—including climate change and its potential reversal.

4.12 Rich and Poor in Science

Bernardo Alberto Houssay (Buenos Aires, 1887–1971) won the Nobel Prize in Medicine in 1947, initiating an important research school in Argentina and Ibero-America. Like other outstanding Latin American scientists, he had to fight against all odds—not without significant difficulties—to pursue his scientific endeavors in his country. He was also profoundly concerned with the underlying causes of these difficulties and committed to implementing procedures to neutralize them.

The personal histories of most great Ibero-American scientists share a common denominator: the development of individual work leading to the highest international recognition in science; the struggle to institutionalize scientific development within their country's cultural sphere; the challenge of convincing political and economic powers of its necessity; and, finally, the progressive dilution of the scientific school they built, often due to the vicissitudes of their country's political history.

This reality is well captured in Houssay's forceful statement: "Rich countries are rich because they devote money to scientific-technological development, and poor countries remain so because they do not. Science is not expensive. Ignorance is expensive."

Only those concerned with science in countries with a limited scientific tradition could write such a sentence. Houssay understood how investment in science and technology has gone hand in hand with the generation of wealth in countries that have embraced it. He recognized that without such promotion, poor countries would remain poor. But what is most relevant, in my opinion, is his assertion that science is not expensive. Instead, what is costly is not investing in it, because leaving a country ignorant of science is tantamount to impoverishing it.

The cost to a country of acquiring products, patents, or equipment from nations that have developed them through science and technology is a direct cause of impoverishment. Scientific and technological dependence is costly. A country cannot produce all types of science and technology, but it can make enough to offset the wealth it spends on acquiring foreign products. This is the justification for Houssay's claim that science is not expensive. One only needs to compare the cost of dependence with the potential investment in domestic science to understand the point. The proverb applies: "Bread for today, hunger for tomorrow."

Where, then, does the problem lie, if it is as simple as investing a reasonable and justifiable amount of public and private resources in science? The answer is linked to another question: Why do those who can invest in science choose not to? Is it because they do not see its value? Because they prioritize other urgent matters? Or because they know it won't bring them political gain?

Education and health, for example, are two fundamental rights; we do not question whether a large part of the budget should be allocated to them. But science should be incorporated as a third fundamental right of citizenship—a notion already being discussed in many international forums. The political class attends—or should attend—to the first two without hesitation; they are not a business, or at least should not be. But why not incorporate science as a third right, once and for all, in our countries?

Doing so would likely eliminate the doubt politicians face when considering investment in something with few short-term, tangible returns. As I've noted before, politicians can gain electoral support by satisfying citizens' needs in education and health. However, demonstrating the long-term benefits of investing in science and technology is significantly more challenging. I have already emphasized the importance of science education for citizens, and how this can encourage the political establishment to invest in science and technology in our countries.

4.13 Realism, Science, and Climate Change

Realism is a philosophical doctrine that emphasizes the objective existence of universal concepts. It is a form of knowledge based on the idea that objects perceived through the senses have an existence independent of perception. Without delving into further philosophical disquisitions, let us assume that science feels comfortable under the umbrella of this doctrine.

In a reflective article entitled "*Climate Change and the Extinction of Thought,*" published in *El País* on June 9, 2019, the well-known essayist John Gray, Emeritus Professor of European Thought at the London School of Economics, paints a bleak picture of the future of our species in relation to climate change. Echoing what science has long demonstrated—that our species is depleting resources and disrupting the planet's complex balances through increasing population density and a wide range of actions—Gray anticipates a grim future.

Although he claims to rely on realistic thinking and science, Gray urges us to abandon the deeply rooted belief that our species will overcome this problem, as it has many others throughout history. For Gray, the problem already exists, and its effects cannot be alleviated in the short or medium term. He therefore proposes political actions aimed at saving what can still be saved—a sustainable retreat that does not compromise economic progress, since slowing it down could lead to geopolitical destabilization and conflict. In short, a very bleak outlook. I get the impression that Gray distrusts his own species, and although he argues that we willingly accept what science tells us, I'm not entirely convinced that he believes it himself, as he seems to lack faith in those who do.

However, I present a different perspective here. At the graduation ceremony for MIT (Massachusetts Institute of Technology) students on June 8, 2019, Michael Bloomberg—a well-known businessman, philanthropist, and three-time mayor of New York City—delivered the commencement address. He spoke to the graduating students about the relationship he perceived between two pivotal moments in human history: 1969 and 2019.

In 1969, under the legacy of the late U.S. President J.F. Kennedy, who had championed the space program, NASA's Apollo 11 mission successfully landed humans on the Moon. MIT played a significant role in the engineering required for this achievement. To a large extent, the goal was not merely to set foot on the Moon, but to catalyze technological development, particularly in the field of communications. Indeed, that program accelerated global technological progress.

Fifty years later, Bloomberg identified a new target: climate change. On a national note, he acknowledged that "our current president" was dismissive of the issue. In 1969, geopolitics revolved around space exploration, pitting the U.S. against the former Soviet Union. Today, climate change is a geopolitical issue—one that no country can or should ignore. Bloomberg argued that denial is no longer an option. Science has already demonstrated the reality of climate change beyond question. And because of its extraordinary complexity, the actions and solutions it requires will also be outstanding and multifaceted. That was his brave and hopeful message to the graduating MIT students: Destination: Climate Change.

As you may have noticed, these are two radically different messages. Fatalism and optimism—both grounded in science, although, as I've mentioned, I have my doubts about the degree of agency Gray attributes to science. These two attitudes reflect the urgency and complexity of a challenge that, perhaps more than any other, demands all our intelligence and willpower.

4.14 The Pact Between Technology and Humanity

Humans are the only species that can see their history, prehistory, and connection with the rest of the world. It would seem that we are, with high probability, the only species capable of directing its future. From this perspective—of being able to redirect our future—I have argued that, in the hands of science and technology, beyond what biology itself imposes, we could well speak of transcending ourselves as a species and witnessing transformations in ourselves, giving rise to other entities with a new ontological status, as well as similar transformations in other living beings. Transhumanism and transhuman evolution are terms that describe these types of transformations.

However, I believe it is also necessary to emphasize the urgent need to consider the nature and consequences of the activities mentioned above. It is essential to highlight this urgency because it underscores the importance of humans thinking critically. And it is here that I perceive something is not working as it should—as if science itself and, more particularly, technology are in a nascent form, acting on the generality of the species in such a way that they are placing us below our capacity to discern and analyze critically.

When I say that science or technology is leading us to this situation, what underlies it is the action of the large corporations that manage digital technologies and the Internet of Things, which drag the ordinary mortal into a kind of "unconditional," uncritical acceptance of the supposed benefits of the digitalization of the world, including ourselves. The process is very subtle; we hardly notice how our lives change and how, without realizing it, we gradually fall into the hands of and become dependent on this whole new world, which, in reality, has already changed everything.

José María Lassalle (2019) has described it beautifully in his book *Ciberleviatán. El colapso de la democracia liberal frente a la revolución digital*. Lassalle raises his critical voice against the unabashed promotion and ideological defense offered by digital corporations—most of them located in the United States and China—who sell us a vision of a wonderful new world that awaits us, one that will arrive when artificial intelligence, robotization, or the most advanced internet offer us better solutions for everything in our lives, from everyday problems to those of greater magnitude. They are tenacious visionaries who abhor rules and regulations—arguing that they are bureaucratic and far removed from the spirit of entrepreneurship and creativity that should govern the species' destiny, if only for "the rest of humanity." Still, the truth is that their technological products are also for that great rest.

In reality, let's not fool ourselves. We are already immersed in this world, and I wouldn't know if we are comfortable with it. We are always waiting for what is to come, which will supposedly be better than what we already have. The digital is becoming a soma that abducts and infantilizes us to the point of not only creating a vital dependence but also preparing us, when the time comes, to renounce our inalienable rights to freedom, democracy, and the exercise of criticism.

This is not the transhuman or transevolutionary world I envisioned. But I must admit that I may be guilty of naivety if I do not recognize that we are moving toward a kind of infantilization that is placing us further below our capabilities. In response to this, very concrete actions are required—actions that imply a profound reflection on the digitalization of the world. Faced with the growing wave of digital technologization we are experiencing, we will need a technology pact sooner rather than later. Let us not forget that there are well-known "gurus" who advocate digital technology as the solution to all the problems that beset us, without seeming to care about what it means in terms of altering our ontological status—something like: "Leave us alone, we will provide you with the happiness you have dreamed of."

I identify with Lassalle's proposal, which calls for the involvement and repositioning of Europe and Latin America in the face of the Calvinist Cyber Leviathan that is approaching. Despite all the problems the peoples of these continents have faced and continue to face, they nevertheless hold—through their histories and philosophies—fundamental values of democracy, freedom, and the recognition of the human being as an inalienable subject. These values position them exceptionally to advocate for a global pact between humanity and technology.

4.15 In the Hands of the "Algorithm"

A summer thunderstorm lasting only a few minutes over the Boston airport was enough for the plane coming from Madrid—which I was supposed to take back to Spain—not to land there and instead be diverted to one of the New York airports. After a few hours, the flight arrived in Boston, but it was already too late. The crew of the return flight had been waiting as long as the passengers, but that wait, combined with the duration of the return flight, exceeded the maximum working hours allowed. The commander, I assume, with the company's approval, canceled the flight after six hours of uncertain waiting in the boarding lounge.

I had to arrange my return and find a new flight as soon as possible. There didn't seem to be much of a problem. I called the company, and they were very kind in providing one. A few hours later, I tried to obtain the boarding passes through the company's website, but to no avail. I contacted the company again by phone, and they confirmed that, although everything was in order and I had a ticket and assigned seats, they could not explain why I couldn't get the boarding passes. They confirmed that "the system" didn't know why (this issue occurred quite frequently) and didn't respond adequately when a case like mine arose.

Couldn't they give me a reason? It's easy to conclude that someone could—at the very least—be the one who developed the algorithm that issues boarding passes. And I had to accept, as I suppose everyone else does, that the people assisting me were neither aware of nor adequately trained or instructed to explain what was happening, let alone how to fix it.

An enormous uneasiness came over me after hearing the comments of the people assisting me, acknowledging that the system "behaved" like that from time to time—that sometimes, within hours, it would be resolved, but often not—and that I shouldn't worry because, in any case, the situation was under control. I was registered in "the system," and my boarding pass would be issued, of course, at the company's counter, because the "system" there would not fail. But why should I resist thinking that one day the "system" could collapse altogether—at least for me, and even more so for everyone?

The "system"—that particular algorithm and any other—has become just another part of us or, probably, a much greater part of our lives than any of us. We can no longer do without them; if the system fails, crashes, or stops, it can affect our ability to act. The idea that algorithms are increasingly able to do everything—supposedly much more efficiently than we can—in reality becomes a dehumanizing drug. Algorithms place us below our potential for action by granting them capabilities—accepted uncritically—which, being superior to ours, also make it easier for us to lose our own through lack of use, something like the progressive promotion of stupidity.

But the algorithm is not futuristic; we can no longer live without them—or, in other words, the effort required to counteract the enormous convenience they offer is considerable. This is a double-edged sword: we must be prepared and educated to know what we can and cannot entrust to their control. It is incredible to think we can live in a fully algorithmic world, even if the algorithm is unconscious.

We speculate about making the algorithmic machine conscious. However, is it not more necessary to reflect on what is already happening with algorithms that are not conscious but immense in their capabilities, than on humanity in a world where that singularity has been achieved? In other words, let us not stop thinking about what is already happening because we want to focus exclusively on what conscious algorithms might do to dominate the world. What unspoken interests may be behind all this, and who are the beneficiaries?

4.16 The Logic of Science and the Logic of Nations

I have always considered that we go hand in hand with science—that it is our best ally, the rational activity that distributes its benefits most equitably, and the one we can rely on most when facing the significant challenges ahead. However, I would also argue that we must acknowledge that some denigrate science, viewing it as a new institution allied with certain public and private powers, and turning it into a new religion that imparts blessings and benefits to those specific powers.

The thesis—which I share—that "science is universal and infinite" contains more than one problem when considered as a social institution. A useful parallel to illustrate my point is that of the Church. Spirituality or faith can be described as universal and infinite, something that can reach everyone, even if not everyone is ready to accept it. However, the Church, as a social institution, has always aligned itself with power—if not by becoming power itself.

It can be accepted that the scientific explanation of the world is as legitimate an aspiration of science as spirituality or belief in a supreme being is for giving meaning to existence. However, both have their undeniable institutional aspects, with distinct trajectories throughout history. I believe this aspect is in the minds of science's detractors, who link the results of its activity to public and private interests.

I would particularly like to emphasize the question of "interests," because there is no longer any national or international research plan or program that does not contemplate—or make explicit—the idea of making the products of science economically viable. This term is not typically used, but the improvement of the quality of life for citizens of the countries involved is always kept in mind. The two go hand in hand: progress is made in knowledge and explanation, while simultaneously, the return on investment in science funding by political and financial powers is considered or desired. On the other hand, the interests of the private sector in science are contrasted with those of the public sector, suggesting that the latter serves the interests of all citizens. In contrast, the former serves only investors or shareholders. We forget that the public sector also has its own "investors and shareholders."

Science financed by the public sector is linked to countries or alliances between countries. The return expected by public authorities when they fund a particular scientific endeavor is—or is believed to be—primarily for the benefit of the citizens of those countries. When we hear that certain countries invest more in science and innovation than others, the assumption is that this investment will improve the quality of life of their citizens. Therefore, each country, in exercising its political and economic autonomy, makes its citizens the de facto investors and shareholders in relation to those of other countries. It is curious that, although the claim that science should benefit everyone is loudly proclaimed, the fine print reveals that the most direct beneficiaries will be a select few.

I do not mean to imply that each country's public sector is unwilling to pursue the supreme goal of universal science—that it should benefit all. It is said that the legitimate interests of countries justify such investments. But it is not in the most profound logic of these actions that such a claim to inclusion is truly made, despite the flamboyant declarations of universality. It is the "legitimate" logic of countries and nations. However, as an institution, science could pursue other interests and detach itself from the powers of states by adopting a different principle of legitimacy. In my opinion, it is very timely to bring this reflection forward, as we can imagine other ways of conceiving public sector financing of science that would be far more inclusive of all humanity.

4.17 Information and Noise

Are we prepared to handle such a large amount of information? The keyword in this question is "so much." We have never stopped receiving information that is somewhat abstract and general. First of all, everything carries information, and that information is always transmitted to us with a certain degree of inevitable degradation. Information theory—a mathematical framework proposed in the 1940s by Claude E. Shannon and Warren Weaver—is relevant here because it lays the foundation for understanding how a message, initiated and encoded at a source, is transmitted through a channel to a destination, where it is decoded and interpreted. The theory explains how initial information is partially lost due to the noise inherent in the transmission process.

Although this is not a metaphor—since it is possible to measure how much initial information from a source reaches us, i.e., our brain, through a channel—it can be a helpful way to grasp the colossal problem we currently face: how much information are we actually capable of processing, given the overwhelming volume that surrounds us? A little reflection reveals that it is not exactly the information itself, but the proliferation of sources, that multiplies before us. In the not-so-distant past, in non-globalized societies, there were fewer sources of information. These had to be known and processed to navigate daily life. But increasing access to a globalized world has placed before us an almost infinite number of "things" whose information we are expected to process. But can we?

There is a physical limit to our brain's capacity to process external information, and it is likely that particular messages from specific sources are not processed at all—meaning those sources effectively do not exist for us. However, what I want to highlight here is not just that issue, but another, equally important one: whether a similar phenomenon of non-processing occurs when the sheer number of sources neutralizes their impact. That is, some sources may generate so much noise that they drown out others, and ultimately, none of them reaches us clearly—or if they do, they are so distorted that the information received no longer corresponds to the original message.

We are not biologically equipped to handle an unlimited increase in information. Unless we develop ways to enhance the brain's processing capacity—through synthetic biology or other forms of cognitive augmentation—it matters little that the sources of information continue to grow. We have a limit. But I would also argue that we are not culturally prepared to receive increasing amounts of information. Education, in this respect, should teach us how to navigate a world with "so much" information. What underlies this thesis is the urgent need for a robust program of civic education in response to the relentless flood of stimuli presented to us in a globalized world. If we are not equipped to filter information, we run the serious risk of concluding that "anything goes."

And what about the use of this information? In light of the above, we face a pressing problem—one that should concern all authorities, governments, institutions, companies, and others that release products and messages to the public without considering the impact on others. Suppose the noise created by others distorts the messages we receive from these sources. In that case, there is a significant possibility that the final information received does not correspond to what was initially sent. This situation paints a rather pessimistic picture of how information is used. Even if all these sources provide accurate and honest information—despite contradicting one another—the result may be interpretations that are incorrect and disconnected from reality. All this, purely because we receive too much information—something to think about.

4.18 To Each His Own

I have always defended the idea of the universality of science. Contrary to what is often cited as an example of this—that the world's leading research centers attract the best minds on the planet to develop it—my commitment to universality goes in another, though complementary, direction: that science can be created anywhere. You might disagree with this thesis, as it likely implies an unaffordable economic effort by the governments and private sectors of nations where science lacks a strong presence. My argument in favor is that this effort should be proportionate and directed toward promoting the development of feasible science in each place.

An additional element related to the claim of science's universality is the consideration of what science has accomplished throughout history, since it emerged in its modern form in the sixteenth century. And when I say "what science has accomplished," I mean, again, science at any time and in any place. This is fertile ground for historians of science, who should uncover what was discovered, when, by whom, and even who may have found it earlier. We risk having a skewed view of scientific breakthroughs as belonging only to the countries that are now the most scientifically advanced, especially if historians from those countries are the ones writing the history. This is not necessarily due to national chauvinism, but rather to the possibility that historians from countries currently less advanced in science have not done—or have not been able to do—that work. And let us not forget: the geography of scientific achievement shifts over time.

Although I am not a historian, I am aware of significant innovations in Ibero-America, such as those in Argentina, Brazil, and Mexico. However, I will illustrate this with three examples from the history of science in Spain that appear to have been rediscovered later by others. First, astronomers and mathematicians at the University of Salamanca in the sixteenth century played a crucial role in establishing the calendar we use today. Second, the treatment of sailors affected by scurvy is often attributed to Royal Navy surgeon James Lind, who documented it in a 1753 publication. Yet in 1579, Fray Agustín de Farfán had already written about the condition, recommending the use of fresh citrus fruits and derivatives aboard ships. Third, and more recently, the invention of the epidural is credited to Spanish military doctor Fidel Pagés, who published his findings in a Spanish surgical journal in 1921. A decade later, in 1931, Italian surgeon Achille Mario Dogliotti rediscovered the technique and was long considered its official discoverer—though he later acknowledged Pagés' prior work.

I offer these three examples to illustrate the complex dynamics surrounding the universality of science and the recognition of scientific discoveries across time and geography. First, time—because the further back we go, the harder it becomes to find proper documentation, not to mention discoveries lost in other cultures and civilizations. Second, geography—because written histories often reflect the efforts of those who have searched only within their own countries, especially those now dominant in science. And third, the matter of restitution—the recognition that someone else had already made the discovery, and the exemplary acceptance of that fact by the scientific community.

Although I consider this practice of recognition very healthy—as it helps temper feelings of superiority among nations—we must not overlook the importance of each country delving into its own scientific historiography. More importantly, we must cross-reference these histories to set the record straight. These reflections are intended to highlight historical and structural asymmetries, not to deny the universal aspirations of scientific knowledge.

4.19 Selecting the Best?

I have seen that what makes two academic institutions—Harvard University and the Massachusetts Institute of Technology (MIT)—great is their students. They admit the best in the world into their classrooms; let's say they can afford to select from among the most outstanding and creative applicants worldwide. And many are left out. We should not be too quick to assume that only a privileged few gain admission due to favorable social and economic status. I'm not saying this doesn't happen, but I know that the profile of those admitted under such conditions is broader than it might initially appear. On the other hand, the selection processes are rigorous. Although high minimum scores are required in a specific set of qualifications corresponding to secondary education, an interview with an admissions committee or a personal statement from the applicant—often a decisive factor in the final admission decision —is also typically required.

It is worth reflecting on this highly selective recruitment phenomenon. Each young person who applies to one of these institutions (and there are many other excellent ones around the world, it must be said) has a unique story. Some individuals possess natural talent and have received an outstanding education, while others have not. We already know that nature, culture, and education work together to foster the emergence of talented, creative, and critically minded individuals. For now, I won't address the biological basis of talent, but let's assume that each person is born with certain gifts. Much more can be said, however, about education and culture as essential components of talent development—areas where social intervention is a reality. Whether we like it or not, we acquire culture and educate ourselves, whether it's for better or worse. Here, everyone has the right to demand that education and culture be designed to foster talent, creativity, and critical thinking.

By no means do I suggest that we should all become the same. Human beings are unique—this is intrinsic to life: no two entities are identical, not even two clones once they've developed. Therefore, I advocate for the development of that uniqueness so that each person can realize their full potential. I have often reflected on how much talent has gone unrealized, how much has been wasted, and how humanity has missed out on it for its own advancement. Under this broader conception—that education and culture are fundamental goods that must be instilled in children and young people—I am not only invoking the right recognized in the Universal Declaration of Human Rights, but also the right to education, as enshrined in Article 26 of the International Covenant on Economic, Social and Cultural Rights. I am also imagining what humanity might have been, what our civilization could be today, had we nurtured more talent, creativity, and critical thinking on average.

Can we measure this? We can assess whether education and culture are improving in different countries, but we cannot rely solely on average values per country or globally. We need to determine whether we are closer to where we ought to be—and whether everyone is receiving the education and cultural exposure necessary to maximize their individual talents. In other words, we are dealing with a pure utopia: Are we all as good as we could be? We are not. There is still a long road ahead—socially and politically—before we can affirm that the education and culture received by people today, and in the past, are optimal. I believe it is feasible, however. But we can already answer the original question: institutions do not select the best possible—they choose a subset of the supposed best at a given moment. There is still a long way to go.

4.20 Right to Scientific Knowledge

Nothing is more liberating than education and culture. They are tools that give you the credentials to be autonomous and critical. You may live in circumstances where the criticism to which you are subject—everything around you—makes you see that you are deprived of freedom or subjected to forces beyond your capacity to act. But that will be a conclusion you reach. If you lack adequate education and culture, you could live without knowing where you live—passing through this world merely to survive, submerged in ignorance.

I begin with this reflection to underscore the importance of this individual right. The deprivation of this right—whether voluntary or imposed—by any country, institution, group, or individual upon any citizen of the world is a true crime of lèse-majesté. Framed this way, we can already see how far we still have to go to make this right a planetary reality.

Science is part of lifelong education, and cultural instruction is essential for every individual. As I wish to emphasize here, science is a citizen's right—perhaps with a more universal scope than any other form of knowledge. At its core, science impacts everyone through its effects. It may be objected that this has never truly been the case. However, I argue that science carries a moral imperative to reach everyone through the way it explains and acts upon the world. If it fails to do so, it is due to a different, more confusing issue: if not properly thought through, science is one thing, and the actors of science are another.

Science is a way of knowing the world, and anyone—I repeat—can access it. There is nothing in science that prevents any citizen, anywhere in the world, no matter how remote, from making use of its excellence. Without invoking grand principles or assumptions—though increasingly solid theories about the world's reality have accumulated in its saddlebags—science has helped us understand how things are. And the best part, although it may seem counterintuitive, is that science does not take pride in its achievements—it can always revise an explanation in favor of a better one. Science is subject to constant regeneration. That is why I consider it a fundamental tool for cultivating critical thinking, thus linking back to the importance of education and culture mentioned in the first paragraph. Science must be part of that foundation, though I do not claim it is the only form of knowledge—or the only one that should nourish our education.

How can we help science reach the citizens of the world? Without the arrogant pretension of being the exclusive form of knowledge—a thesis that likely contradicts the advocates of scientism—science is the best traveling companion to help us understand and make decisions. Yes, we—private individuals, then and now, anywhere in the world. Science, in this sense, is a great little right that must be served in small sips and continuously, always bearing in mind that the final recipient—I repeat—is any citizen. That citizen must learn it while simultaneously utilizing it. I doubt there is a more precious, more suitable product manufactured by humanity with the potential to reach each one of us than science.

I use the term "potentiality of science" to emphasize its capacity to reach everyone—and to show, as with many other rights not yet universally realized, that the road ahead is still long before we achieve the actualization of that potential. However, science is a double-edged sword. Even if its realization were to come one day—and all citizens were educated in it—the reality it presents to us might not be to our liking. That is the first edge. The second is that we would, at least, have the critical capacity to confront it.

4.21 Climate Change, Science, and Multilateralism

Disappointingly, we read that the UN climate summit—the 25th, which has just concluded in Madrid —failed to produce a strong commitment from countries to take decisive measures to prevent global warming. I will not delve into the details of what such measures should entail. Instead, I will focus on another matter. Only eighty-four of the approximately two hundred countries present were willing to propose strict actions against the greenhouse effect in 2020. The majority have not done so, including those countries that contribute most to it.

We are faced with three issues. The first is the global problem itself: climate change and the well-documented trend of rising average temperatures resulting from carbon dioxide emissions—the primary greenhouse gas. The year 2019 set a temperature record, continuing an upward trajectory over the past decade.

The second issue is multilateralism. The problem belongs to all of us—the house is shared—and it seems evident that everyone living in it should contribute to maintaining or repairing it so that it does not collapse. If this seems self-evident, why is it so difficult to reach political consensus or joint action? The answer is also straightforward: nations have historically been configured to develop policies focused on "their" territories and everything that entails—primarily the protection of their economies and the welfare of their citizens. There may be alliances or agreements between nations, even annexations or territorial expansions, but the objective of national leaders is to prioritize the interests of their own territories and populations. The planet itself is an entity that escapes them—almost by definition.

Nations can be asked to commit to climate action, regardless of whether some political leaders deny the reality of climate change. That denial is no longer the central issue. The issue is the unilateral inertia of nations and, consequently, the challenge of multilateralism in achieving standard policies. Because, as I've said, nations will eventually make decisions on climate change—but when? If the urgency of the matter demands unanimous action, we cannot afford to wait for each nation to act according to its own economic timetable.

The third issue concerns the role of science. Scientific communities around the world—working in universities and research institutions, both public and private, commissioned by national and international agencies—have already spoken. Through publications in leading journals and presentations at scientific congresses and interdisciplinary forums, science has established its position: we must act, and we are in a position to recommend the kind of global actions that should be implemented. However, the observations and predictions of science are one thing—their calls for action are another. Because science, ultimately, is a consultant—it is not a binding authority.

As an institution, science lacks executive power. Whether considered at the national level or within supranational entities, it functions as an advisory body. Others make the decisions. You may think this is how it should be—that political leaders should hold the power to make those decisions. However, if, in the face of a crisis as severe as this one, those decisions are not made or not implemented unanimously, what is going wrong?

References

Bermejo D (2019). Pasión por la vida. Passion for life. Alianza Universidad, Madrid
Dessauer F (1965) El caso Galileo y nosotros. Ediciones Carlos Lohlé, Buenos Aires
Gould SJ (1981) The mismeasure of man. W. W. Norton and Company, New York
Heidegger M (1996) Einleitung in die philosophie. Vittorio Klostermann GmbH, Frankfurt am Main
Lassalle JM (2019) Ciberleviatán. El colapso de la democracia liberal frente a la revolución digital. Arpa Editores, Barcelona
Latour B (2019) Dónde aterrizar. Cómo orientarse en política. Penguin Random House, Barcelona
Pinker S (2018) Enlightenment now: the case for reason, science, humanism, and progress. Viking, New York
Ridley M (2010) The rational optimist: how prosperity evolves. Harper Collins, New York

Chapter 5
Natural Evolution

Abstract Here, I reflect on biological evolution as a historical and scientific process that continues to shape our understanding of life. I begin by revisiting Darwin's "dangerous idea"—not only natural selection, but the deeper notion of a shared tree of life connecting all organisms. I explore why this idea still provokes resistance and how modern biology, through genomics, developmental genetics, and systems biology, has expanded and refined Darwin's insights. I examine the genetic and cultural dimensions of human violence, the evolutionary roots of our species, and the coexistence with now-extinct human relatives. I also discuss the role of science in legal contexts, the challenges of teaching evolution, and the philosophical implications of progress in the field of biology. Through these reflections, I argue that understanding evolution is essential not only for science but for how we see ourselves—as beings shaped by contingency, complexity, and a shared biological heritage. The evolutionary processes discussed here are presented as historically contingent and open-ended, rather than as deterministic trajectories.

Keywords Complexity · Contingency · Human violence · Natural selection · Progress · Teaching evolution · Tree of life

5.1 Darwin's Dangerous Idea

It never ceases to amaze me, time and again, the motives that drive the heterogeneous group that opposes the theory of biological evolution. I try to put myself in their shoes, attempting to understand what leads them to this persistent rejection of the evolutionary fact and, logically, its corollary: that our species is just one among many in the tree of life. It can be admitted—since we are operating within the realm of science—that the theory of evolution is subject to refinement, like any other scientific theory. However, with Darwin, the theory gained a solid footing: it seeks to explain a reality—the fact of biological evolution. One might think that Darwin's "dangerous idea," a phrase borrowed from the title of Daniel Dennett's 1995 book, lies in his explanation of evolution through natural selection. However, Darwin's

A. Moya, *Reflections on Science and Evolution*,
https://doi.org/10.1007/978-3-032-25682-9_5

other significant contribution is even more far-reaching and unsettling: the theory of evolution, the material connection among all living beings since life first appeared on Earth.

The reason for my astonishment is complex; several factors contribute to it, which I will now outline. The first is the relative silence of the group mentioned above regarding other theories, such as biological evolution, which are also based on natural explanations of phenomena in the world. The laws of physics—from Galileo to Newton, through Einstein, and culminating in modern cosmological theories about the origin of the universe—offer explanations of reality that are no more comforting to those who rely on non-natural "reasons." One might wonder whether the lack of recent public hostility toward physical theories, compared to the theory of evolution, stems from their greater difficulty of comprehension. However, this justification is unsatisfactory, as fundamental physical theories can be understood without resorting to complex mathematics. Moreover, the best physicists of our time have made immense efforts to disseminate and explain these theories, presenting both the panorama of the universe's origin and the intricate composition of matter, as well as abstruse theories concerning the unification of fundamental forces. Therefore, this first possible point of difference—favoring the acceptability or non-belligerence toward physical theories over evolutionary theory—does not seem to lie in the theories themselves.

I wish to assert that both types of theory form an explanatory continuum within the field of science. If physicists tell us something fundamental about the origin of the universe without invoking the supernatural, evolutionary biologists do the same regarding the origin and development of life, including our own species. From this perspective, anti-evolutionary belligerence should be viewed as a form of belligerence against science itself. There are criticisms of science based on its supposed pan-explanatory arrogance, which come from an intellectual tradition quite different from the group I refer to here. That tradition is part of a secular current in Western history that resists unveiling the ineffable, arguing that scientific explanation fails to grasp the essence of things. This ineffabilist view holds that science cannot capture the true nature of reality.

Darwin was acutely aware of the scope of his dangerous ideas. The deeper motive behind the group I refer to—their undeclared concern—is a reluctance to accept what we now take for granted: that we are just another species, a product of biological evolution. But science knows no boundaries. We are one more species—but are we necessarily just any species?

5.2 Man Is a Wolf to Man, but not too Much

We already know that the human species is violent—and that violence is malleable through education. Like many other traits, violence has both a genetic and an environmental component, and the latter is shaped by a combination of factors, including education. But how much of this violence is innate, intrinsic to our nature?

In the study by Goméz et al. (2016), the authors delve into the genetic, sociological, and political bases of the most extreme form of violence—what they call "lethal violence": violence that occurs between individuals of the same species and results in the death of a fellow human being. Violence is a concept with a long history, as many human behaviors fall under its umbrella. As is always the case in science, providing a precise definition is essential for accurate measurement. The authors, as noted, focus on a specific form of violence—the most extreme—and calculate it as the percentage of deaths caused by violence between individuals of the same species, relative to other causes of death. Lethal violence includes war, homicide, infanticide, execution, or any other form of murder.

This rigorous measurement of lethal violence is based on well-documented sources from around six hundred human populations, spanning periods from the Paleolithic to the modern and contemporary ages. The authors also group populations by socio-political organization—distinguishing, for example, between bands, tribes, and chiefdoms in prehistory and history, as well as populations belonging to modern states.

In parallel, the authors estimate lethal violence in over a thousand mammalian species, including documented cases of infanticide, cannibalism, inter-group aggression, or other forms of killing. These species exhibit a wide range of behaviors, varying widely in terms of sociability and territoriality. Some are evolutionarily close to us—namely, Neanderthals, great apes, and other primates. Like any trait with a genetic basis, lethal violence can be considered subject to evolutionary pressures.

The evolutionary analysis of lethal violence across all mammalian species examined allows the authors to determine the level of lethal violence expected for humans based on our position in the tree of life. The result is striking: 2%. That is, we would expect a 2% rate of lethal violence based on our genetic makeup. And how much violence have we exhibited throughout our prehistory and history? Recall that non-genetic components also influence this trait. These are not inherited like genes, but they can be transmitted and learned—especially in species like ours that have developed learning, communication, and culture. For example, the level of lethal violence during human prehistory, including the Bronze Age, aligns with the 2% expectation. It is higher in populations from the Iron Age and historical periods, such as the Middle Ages, but falls below 2% in populations from the modern and contemporary ages.

Regarding socio-political organization, the authors observe that prehistoric bands and tribes exhibit lethal violence at a rate of around 2%. In contrast, chiefdom societies and contemporary bands and tribes exhibit higher levels of complexity. Populations from other historical periods or modern states, however, show notably lower values, below the 2% threshold.

In 1661, philosopher Thomas Hobbes, in his *Leviathan*, described man as "a wolf to man." We have long believed he was somewhat correct, but science now offers a more conclusive view—at least regarding extreme lethal violence. Indeed, we carry some intrinsic violence; it comes from afar, and we should not forget it. It is part of our nature. It is no longer said that we carry it in our blood, but in our genes.

Yet history shows significant fluctuations due to non-genetic factors that have both increased and decreased violence. Importantly, advanced social organizations—particularly modern states—appear to have contributed to reducing lethal violence below our genetic baseline. Like many other species with similar social and territorial structures, humans have a 2% genetic propensity for lethal violence. However, violence can be reduced—and in fact, it has been—through education, culture, and social organization.

In our most remote past, in prehistory, when such institutions did not exist, we did not live in a happy Arcadia: the lethal violence we exhibited was that which corresponded to our nature. Later, depending on the context, education, culture, civilization, and state structures have contributed to either increasing or decreasing that 2%. In light of our history, there is nothing new to invent or discover about how to reduce this value.

5.3 Those Who Were Also Human

Among the outstanding contributions of science to Western culture, two particularly relevant ones stand out for their challenge to anthropocentrism. The first is Copernicus' demonstration that the Earth is not the center of the universe—it revolves around the Sun. The second is Darwin's formulation that species originate and evolve from one another, and that humans are also the product of that evolution. But what is surprising, at least to me, is that this contribution against anthropocentrism only gained traction nearly three centuries after the Copernican milestone.

The reasons for this delay are manifold, but one stands out as particularly significant: even if the evidence for the non-centrality of the planet that hosts us was irrefutable, nothing could exclude, a priori, the belief that it remained central in the universe because it hosted a being—the only being—whom the Creator had endowed with a soul. To what extent has evolutionary theory contributed to undermining such a conception?

Although evolutionism is an ancient idea, already present in Greek philosophy, we can trace its path throughout Western history. As we approach Darwin, the accumulation of evidence increasingly undermines the credibility of the thesis that our species has a supernatural origin. That evidence, which culminates in Darwin and expands after him, has always been there, but there has been considerable reluctance to read the book of nature through the lens of science.

We should not be surprised by this. We have long held the conviction of being unique on our planet, for at least one fundamental reason: our awareness of our own singular existence. Such exceptionality may have supported the thesis of an insurmountable distance between us and the other beings that inhabit the Earth. Much has been written and reflected on this seemingly obvious conviction: "I think, therefore I am," wrote Descartes. But such a conviction is ultimately unfounded. Rereading nature through the lens of science, we discover new and disturbing surprises that make the path toward connection possible and bring us closer to other living beings—because "others think and, therefore, also exist."

For some time, I participated, with two colleagues, in a program on Spanish National Broadcasting that aired every Thursday evening, where we discussed the latest developments in science while answering listeners' questions. Do you know what topic generated the most interest? The discovery of other species of the genus *Homo*—such as Neanderthals and Denisovans—that coexisted with us for a significant period of time. Paleoanthropologists have uncovered evidence of complex social organization in some of these species, signs of relatively advanced culture, and indications of symbolic thought and religious behavior. Moreover, comparisons of their genomes with ours show that interspecific crossbreeding must have occurred, as the human species, *Homo sapiens sapiens*, carries genetic material from those now-extinct relatives. It is a fascinating story of scientific research—and it is far from over.

Finally, let us reflect on these findings from the last decade. Suppose it is already unsettling to feel that we are unique and alone in the universe. In that case, it is even more perplexing to know that we coexisted with other species of the same genus who, like us, very likely formulated fundamental questions about the meaning of their existence. And yet—they disappeared. Let us not be so arrogant as to think we will not suffer the same fate.

5.4 Science in the Courts

On May 14, 2007, in the city of Valencia (Spain), a court ruled against an anesthesiologist who was found responsible for infecting 275 patients with the hepatitis C virus. On page 207 of the ruling, it is stated that the following facts are considered proven:

- "There are a large number of people whose infection with the hepatitis C virus is attributable to a specific surgical intervention or medical or hospital procedure.
- The only common link between all these individuals and the surgical or hospital interventions is the defendant's role as an anesthesiologist.
- The defendant is a carrier of the hepatitis C virus, and of the same genotype (1a)—a minority or infrequent genotype in Spain—as that found in the infected individuals.
- The only way the defendant could have transmitted the hepatitis C virus to these individuals is via the percutaneous or parenteral route.
- There is evidence that the defendant made irregular use of opiate drugs used in surgical interventions and intensive care services, and that he used them for himself."

On that same page, the judges also stated: "And, as we said before, having concluded that the defendant engaged in the conduct for which he is accused in this case—the percutaneous use of anesthetic material and drugs for himself, sharing their use with patients and infecting them with the hepatitis C virus of which he is a carrier, by also percutaneously inoculating them with drugs contaminated with the virus—to determine the passive subjects or victims of this conduct, the genetic expert evidence must be taken into account."

Dear reader, Professor Fernando González Candelas and I were responsible for providing, together with a large team under our direction, the aforementioned genetic expert evidence, whose investigation—ordered by the judge overseeing the case and financed by the Ministry of Health of the Valencian Community—took us no less than nine years.

The anesthesiologist had been infected with the virus before 1988, and the patients he operated on contracted the virus after that year. Using well-established procedures in the field of biological evolution—and particularly in virology—we were able to demonstrate that the resemblance between the patients' viruses and the anesthesiologist's virus was significantly greater than that of any other person carrying the same virus but unrelated to the outbreak.

Moreover—and this was decisive in the court case—we were able to date, without prior knowledge, the approximate time of surgery and probable infection for each patient. In many cases, we could estimate the likely date of surgery with minimal error. Indeed, if a patient had contracted the virus at a given time due to the anesthesiologist's actions, the virus would mutate over time in that patient, gradually increasing its genetic variability. By sequencing specific regions of the virus genome and applying molecular evolution methodologies, we could estimate the rate at which the virus evolved and, therefore, the likely time of infection based on the genetic variability accumulated.

I will never forget the astonished faces of the lawyers, judges, and other attendees as they witnessed the extraordinary coincidence between the probable date of infection we presented in court and the date of surgery provided by the court clerk immediately afterward. The effect was incredible and decisive: how could it be so accurate about something that had occurred a decade earlier? We had theories and technologies that allowed us to answer questions that would have been impossible to resolve just a few years before the start of our judicial investigation—a triumph of science.

5.5 On Progress

Although they emerged during the Enlightenment in the eighteenth century, those with the attitude or disposition to embrace social transformation were consolidated as "progressives" in the nineteenth century. Progress became a highly connoted term. Under its banner, groups of intellectuals formed societies of friends, gathered in cafés and theaters, and created associations with various iterations of the term. In that century, there was a firm conviction in directional transformation—not necessarily directed by any one individual, but as the product of collective forces that improved society. The term progress implies positive change, an increase in whatever is being measured—such as wealth, social welfare, or non-violence, to name a few, without delving into how it might be measured or empirically verified.

The twentieth century, however, introduced profound doubts about progress—about, in short, the betterment of civil society—having witnessed, among other things, two world wars. Yet important remnants remained, particularly among those who identified as progressives, in contrast to those who favored non-mobilization—in other words, conservatives. Today, it is not easy to define, in ideological or political terms, the differences between conservatives and progressives. However, this dichotomy may still be more helpful in understanding attitudes toward life than the traditional left-right divide, especially when the notion of a center is introduced into the political spectrum.

Although we have moved from the nineteenth-century conviction that progress exists to a more nuanced view—that progressivism is an attitude and struggle to introduce changes aimed at social improvement, understood as a vital disposition—a renewed belief has recently re-emerged. This time, it is supported by a wide range of data suggesting that human society is indeed progressing, under the enduring ideals of the Enlightenment: reason and science (Pinker 2018).

But what about biological progress? Does it exist? Just as the existence of net social progress remains debatable—despite the evidence mentioned—we can also question the notion of biological progress or improvement. As with social progress, biological progress may be better understood as an attitude or orientation toward life. Darwin saw progress in evolution, as did Huxley, and more recently, Ayala and Dawkins. In contrast, my admired Gould (1997) argued that biological progress is nothing more than anthropocentric prejudice—a projection of nineteenth-century progressivist ideals onto biological evolution, suggesting that if humanity represents progress, then evolution must have produced it as such. Pure ideology, says Gould. Any apparent directionality should be understood as an empirical pattern subject to revision, not as a necessary outcome of evolution.

Gould, however, does not resolve the issue of progress from a scientific standpoint; he critiques it as an anthropocentric ideology. But if progress exists—if it is intrinsic to the evolution of the world, both biological and cultural—then nothing prevents us from imagining that this product of biological evolution we call human can go beyond himself. Some data seem to support this. If there is no ceiling or limit to biological progress, just as there may be none to social improvement, then the advancement of both—whatever form it takes—could lead us beyond our biology and our culture.

In summary, we still lack much more science to grapple with that devilish question: what is progress?

5.6 Teaching Evolution

I greatly enjoyed introducing my Bioinformatics master's students to the theory of evolution. Mine is a non-compulsory subject, and only students who have studied degrees not related to life sciences—mainly mathematics, physics, various engineering disciplines, or those who have not taken courses in biological evolution—are usually the ones who enroll in my course. We had a wonderful time, because through a series of questions I posed or the presentation of specific illustrative examples, I realized that I opened up a whole new world for them, showing how the biodiversity of the planet, past and present, can be explained coherently and scientifically. Probably due to my experience teaching evolution, I know almost instinctively where the core ideas to convey are found—always well seasoned with information from a few decisive experiments or well-chosen results from investigations into the history of life at all levels of organization.

It is precisely through the teaching of evolution that one comes to understand the truth of what zoologist Ernst Mayr (2004) maintained: biology is an autonomous discipline. Autonomous in relation to what? His thesis arises in response to the view of physics as the prototype of scientific rigor. Mayr, and later many others—particularly philosophers of biology—recognized specific components of biology that distinguish it from physics, without diminishing its scientific validity. I do not think this has been easy to demonstrate to the broader scientific community. It is precisely these components that I try to explain, and which, on the other hand, are so captivating to students, especially those coming from non-biological sciences.

I will illustrate this with two examples: phylogenetic analysis and population thinking. The study of the tree of life through phylogenetic methods—the possibility of reconstructing evolutionary history—is not only a fundamental tool of evolutionary science but also a defining hallmark of biology itself. The Tree of Life is an iconic representation of the living world, often used by professors to illustrate relationships among organisms or their traits. The starting point here is the fact of evolution, but it never ceases to amaze me how powerful and intuitive it is to link entities through such an elegant representation.

The other great tool, with a distinct character that so well defines the living, is what Mayr called "population thinking" as opposed to "typological thinking." Thinking about evolution from the perspective of changing populations—and having the conceptual and mathematical tools to do so—brings us much closer to understanding life than when we consider biological entities as defective individuals striving toward an ideal type supposedly defined by evolution. Transcending this Platonism of ideal forms is crucial to understanding how evolution enables transitions from one organism to another, regardless of how different they may initially appear. Population thinking is Darwinian; typological thinking, irrespective of its utility in systematics, is pre-Darwinian and pre-evolutionary.

The study of evolution can be presented through controversies between schools and traditions—an excellent strategy for teaching science. Science is dynamic, and the teaching of biological evolution is no exception. It allows us to clearly see how theories have clashed, both before and after Darwin, how integrative syntheses have emerged to provide a more complete vision of the evolutionary process, and how unexplored areas and open questions still demand new or better explanations.

5.7 More Animals than Gods?

Suppose we place on one side of a scale the part that corresponds to our "animality" and on the other the part that corresponds to our "spirituality"—to summarize in one word our supposed special status in the universe—and we follow the movement of that scale throughout Western history. In that case, we immediately notice a permanent oscillation, with phases where the scale tilts in favor of spirituality and others where it tilts toward animality. It seems that what is reasonable—a balance between both sides—is consistently elusive. But of course, is it not somewhat artificial to arrange the animal on one side and the spiritual on the other? Who can justify that the latter is not a product of the former, and that they cannot be arranged as interconnected parts?

Furthermore, the metaphor of the balance between animality and spirituality implicitly suggests that the right thing to do is to strike a balance between the two. We must examine in detail the history of human culture and the moment when science—and its explanations of the issue at hand—emerges, to realize that we have long believed this balance exists, that these parts exist, and that they are distinct and separable.

For historian Yuval Noah Harari (2016), there seems to be no doubt that science has been decisive in affirming our animality. Darwin initiates—or at least facilitates—the thesis that, as one more product of evolution, we cannot deny our animal nature. This is the famous question of the naturalization of our species. When the study of evolution is presented historically, it is often noted that before Darwin, humans were placed far above other animals on the scale of beings. With Darwin, this view is dismantled, and we are positioned as one among many living organisms.

But what no one could have imagined—or perhaps some did—is that, gradually, it could be argued that everything in us is animality, more or less elaborated. Harari asserts that we are purely sophisticated "biochemical algorithms." Although this assertion goes beyond what science can strictly demonstrate at present, it is through science that we have arrived at this thesis. In short, the balance has shifted toward animality.

Yet Harari leaves a loophole, an open window to reconsider what, for now, still seems to place us "above our animality." I have little doubt that we have arrived here as a consequence of the material evolution of life, and that evolution has produced in us a process still to be fully unraveled—what we call consciousness. Many new-wave humanists, who can no longer deny our animal nature, appeal to consciousness—a fundamental requisite for the existence of freedom in our actions—as a distinguishing factor from other animals.

There are even those who argue—not only humanists like Nagel (2012), but scientists such as Penrose (1989)—that consciousness is not computable and lies beyond the realm of algorithms. Intelligent entities may emerge as products of our activity, even knowledgeable ones, far surpassing any human who has lived or will live. But they will not be conscious entities, according to Harari and others, even though they may come to dominate us. However, authors such as Kurzweil (2006) argue that they will be conscious, which presents us with numerous challenges.

On the other hand, transhumanism—as advocated by Bostrom (2014), and as I have argued for some time (Moya 2009)—places greater emphasis on our capacity to transform ourselves and control the machines we create. This leads us once again to a shift in the balance toward spirituality, now understood and made intelligible through science. Here, I allow myself to maintain that we will reach a proper balance between animality and spirituality—precisely at the moment when we can control our future, or in other words, become masters of our destiny.

5.8 Parasites, Let's Learn from Them

A parasite is an organism that lives at the expense of another organism of a different species, feeding on it and depleting its resources without killing it. This is the general definition given by the Royal Academy of the Spanish Language. As far as some members of our species are concerned, it is said that they are those people who live at the expense of others. This meaning is not wrong at all. It makes sense—because who falls into the category of "others"? Suppose they are other people of our species, which is the most common interpretation. In that case, we are somewhat departing from the canonical definition that a parasite is a parasite of or on another species. We have extrapolated the biological term to our social world to speak of a predictably abundant fauna: those who live at the expense of others of our species.

We must ask ourselves: to what extent are we not parasites? The answer is simple. In my relations with others, I cannot be a parasite if I am offering something in return, if I do not live exclusively at the expense of others, because I also contribute something to them. However, the question is not so easy, because since it is impossible for a human not to live at least partially at the expense of others, we must somehow quantify not only how much we take from others but also how much we offer in return, to conclude whether one is or is not a parasite. This exercise is feasible and has likely already been addressed in human sociology and economics.

But I have not finished with human parasitism, because we can also consider to what extent we are parasites of other animal and plant species. The idea that a parasite is defined by not killing what it parasitizes is undoubtedly attractive, because we generally feed on animals and plants that we ultimately sacrifice one after another. It is also true that we deplete many other living beings of different species. We use their properties or capacities without sacrificing them, living significantly at their expense—although they also live at ours, even because we procure their subsistence. It cannot be denied that there is a power dynamic in the relationship between humans and other species. In my opinion, this relationship extends beyond what is strictly speaking parasitism.

The biological meaning of parasitism generates less ambiguity than when we metaphorically extrapolate it to our species or our relationships with others.

Parasitism generally entails the idea that the parasite lives, at some point in its life, in or on the organism it parasitizes. Moreover, its use of the host is much greater than any benefit the host may derive from harboring it. But evolution is wise, and it is not strategic for a parasite to wipe out its host. The best strategy—which is often observed in the evolution of life—is that there comes a time when there is reasonable coexistence, something like "it hurts, but doesn't kill."

It was once said that parasites evolved by losing many of the characteristics that non-parasitic species exhibit to survive. In reality, this is not the case. Parasites have developed sophisticated mechanisms to survive on their hosts. These mechanisms range from chemical signals to attract other parasites to the control of host behaviors and even sexual behavior. The more evolved protozoan and nematode parasites are notoriously difficult to eradicate or combat.

On the other hand, it must be made patently clear that life is full of parasites. As Carl Zimmer (2001) rightly comments in his award-winning popular science book *Parasite Rex: Inside the Bizarre World of Nature's Most Dangerous Creatures*, parasites are everywhere and exceptionally abundant. Their absence in populations—including those of our species—and ecosystems should not be interpreted as a sign of health. Quite the contrary: something is not working well when the most abundant beings on the planet are not present where they should be. Although it may sound like an oxymoron, there are never too many parasites; it is better to learn from them to control them than to think only of destroying them. Let us think about it.

5.9 The Arrow of Time and Us

There seems to be no doubt that time is one of the great enigmas of science. The Italian physicist Carlo Rovelli, a specialist in quantum gravity, in his excellent reflective work *The Order of Time* (2018), reveals several ideas that continue to surprise us for the radical shift they imply in fundamental conceptions of modern science. Indeed, modern science, founded with Newton—whom Rovelli frequently references—broadly seeks to objectify and rationally explain the reality of things in the world, in the universe, and to make them independent of our perception. Our perception of things and what they are can, in fact, be very different. Or rather, we run the risk of describing what things are through the prism of our subjectivity, such that it could well be that each thing appears differently depending on who contemplates or analyzes it.

Galileo insisted on distinguishing between the primary qualities of things—their fundamental properties, such as weight, mass, velocity, and acceleration—and the secondary qualities, which encompass our subjective perceptions of them.

To avoid subjectivity, it was long considered that bodies exist in space and time, two absolute dimensions entirely external to us and, in general, to any observer. It took Einstein—Rovelli's other great intellectual hero to show us that there are no absolute dimensions, that space and time are relative to the observer, and that they, along with the primary qualities mentioned earlier, change depending on the relative position from which they are considered. With the appropriate caveats imposed by scientific progress, this marks a return to the thought of some ancient Greeks, who gave nature a subjectivist interpretation: "Tell me where you are, and I will tell you why what you perceive differs from what I perceive, who am in another place or time."

However, the most surprising aspect of Rovelli's text is his reinterpretation of time, culminating in a proposal that time is nothing more than our perception of the flow of events in the direction of increasing entropy. Here again, Rovelli offers a unique perspective. I particularly appreciate his statement: "The whole difference between the past and the future can be reduced to the simple fact that in the past, the world's entropy was low. And why was entropy low in the past? In this chapter, I raise an idea for a possible answer, if attention is to be paid to my answer to this question and its perhaps fanciful assumption. I'm not sure it's the right answer, but it's an idea I've fallen in love with. And it could clarify a lot of things."

We have all studied—or should have studied—the laws of thermodynamics, and we know, in particular, the universe's tendency toward disorder. But why toward disorder? The great physicist Boltzmann dared to formulate the idea that particles, atoms, and molecules are in constant thermal agitation. Now, we are not able to know all the possible configurations of this microscopic world because we only interact with a portion of it and with a set of associated macroscopic variables—but not with all possible ones. It is this lack of knowledge that causes our unfocused view, and entropy depends on—or consists precisely in—this unfocusing. That is the core of Boltzmann's theory.

For Rovelli, the initial entropy of the world seems very low to us. But perhaps it is not necessarily so—it's just that we are unfocused, interacting only with part of the macroscopic variables through which we describe the world. Rovelli's fascinating thesis is that the arrow of time may be a product of our perception, rather than a fundamental property of the universe.

5.10 Convergences

I am dedicated, by vocation, to the study, teaching, and research of evolutionary biology. And I am never tired of studying it in either of its two facets. I always find new angles and approaches when I teach it; it continues to surprise me. But above all, it is the remarkable freshness of the latest research I read about biological evolution that leaves me most perplexed and excited. Even though evolution is a relatively old theory, we are constantly uncovering new findings—both theoretical and experimental, as well as historical—thanks to the latest advancements in scientific technology.

A study published in the *Proceedings of the National Academy of Sciences of the USA* (Young et al. 2019) shows how males of vertebrate species that are phylogenetically very distant—namely, a California mouse, a meadow vole, a bird known as the alpine accentor, a poison frog, and a cichlid fish from Lake Tanganyika—and that are monogamous, share a set of genes whose expression in the brain appears to underlie such behavior. Of course, it is pretty typical of evolutionary analysis to observe that males of vertebrate species closely related to the five mentioned species exhibit polygamous behavior and do not show the same gene expression profiles. It is precisely through such comparative analysis that the authors conclude the existence of a set of genes that predispose males to monogamous behavior. The study does not yet provide the molecular mechanisms that ultimately trigger the accumulation of certain hormones responsible for a sustained level of pleasure in males who remain monogamous. But that seems to be the case.

This is yet another example among the many that appear throughout the history of evolution, illustrating the independent emergence of complex traits. Take flight, for instance. Not only do birds fly—though not all of them, as some like the ostrich have lost this ability—but insects and certain mammals also fly. Convergent evolution is a recurring phenomenon in biology: traits are independently developed by evolutionarily distant organisms. These traits do not arise because they were inherited from a common ancestor, but because they were achieved separately.

Convergences are a recurring theme in biological evolution, and more than one scientist argues that they have been underappreciated, possibly containing a hidden message yet to be uncovered. In a purely fictional exercise—almost like the mental experiments Einstein frequently used to develop his theories—I enjoy posing to students the question of whether intelligence might have evolved in dinosaurs, had they not gone extinct, to a level comparable to that of our species. I like to think so, just as monogamy, vision, and flight have evolved independently in species that have survived. And there is no one orchestrating this symphony of life. Of course, an orchestra is an organized ensemble; someone provides input and output, guiding the score and the movements of the musicians. To say that life is such an orchestra almost seems like an oxymoron. Yet, when we observe this complex trait, we are drawn to the beautiful metaphor of the orchestra of life. Convergences suggest pathways, a certain directionality without a conductor, a certain finality without a defined end. They deserve more attention—science, always more science.

5.11 Symbionts

When we read the term "symbiosis," the first idea that comes to mind is that something positive occurs between the members involved. According to the *Diccionario de la Real Academia Española*, symbiosis is defined as the "association of animal or plant individuals of different species, especially if the symbionts benefit from living together." This is not an incorrect definition, although it should be updated, given that the most common associations are not between animals and animals, animals and plants, or plants and animals, but rather between animals or plants and microorganisms. The definition omits the microbial world, but it can be easily reformulated to include it. Thus, symbiosis becomes the "association of individuals of different species, especially if the symbionts benefit from living together." On the other hand, the notion of "benefiting from life in common" is quite accurate. This is where most people generally agree, assuming that if a relationship is symbiotic, it benefits the members of the association.

The term's history is fascinating, and I will offer a few observations. First, symbiosis involves a very close, often long-term relationship. Symbiotic species live together for a significant portion of their lives. In some cases, they become so inseparable that each offspring of the symbionts re-establishes the association in the next generation. The second observation is equally important: the symbionts mutually benefit from the relationship. They provide something to each other that makes them essential to one another. Note—and this is my third observation—that I am deliberately avoiding the idea that symbiosis is always between just two organisms. While that may be the most common scenario, more than two organisms can be involved in a symbiotic relationship.

I raise two further considerations for reflection. One concerns how symbiosis can be evolutionarily successful, and the other relates to its importance compared to parasitism. If one of the central tenets of evolutionary biology is that organisms strive to perpetuate themselves by leaving as many descendants as possible—at any cost—it might seem that the benefit one organism provides to another could come at the expense of its own offspring. In other words, if organisms have limited resources, those used for activities such as symbiosis might compromise those available for reproduction. However, there is no reason to assume that if one member of the symbiosis offers a benefit, the others do not reciprocate with something of equal evolutionary value. In fact, it could be that mutual benefit leads to an increase in reproductive success for all members involved. Although these ideas are easy to formulate, they are not so easy to demonstrate in the real-world dynamics of coexistence. This is currently a very active area of scientific research.

I have previously discussed the term that stands in contrast to symbiosis: "parasitism." There, I noted that the world of living beings is full of parasites—organisms that live at the expense of others. I also emphasized the role of parasitism as a driver of evolutionary change, prompting host organisms to adapt to their parasites. But what about symbiosis? Is the world equally full of symbionts—organisms that benefit one another? I would be particularly interested in knowing the relative prevalence of symbiosis versus parasitism in biological relationships, and above all, which of the two has contributed more fundamentally to the evolution of life on Earth.

5.12 Harvard Teachers

In the first semester of 2019, I spent a sabbatical at the Department of Organismic and Evolutionary Biology and the Museum of Comparative Zoology at Harvard University. This prestigious department has hosted—and continues to host—some of the most influential figures in the field of evolutionary biology. Among them are Ernst Mayr, Stephen J. Gould, Richard Lewontin, and Edward O. Wilson, four scientists whose work I have followed closely due to the alignment between their scientific and philosophical interests and my own. In fact, I have shaped much of my thinking by studying their respective contributions. I must admit I was deeply moved to be working and walking in the same spaces once occupied by these giants of evolutionary science.

Here, I offer a brief reflection on each of them, highlighting what has stayed with me after years of reading and studying their work. While all four made numerous contributions, I will focus on the ideas that have had the most profound influence on me.

From E. Mayr, I retain his explanation of the difference between population thinking and typological thinking. Evolution must be understood in terms of evolving populations, not idealized archetypes. Organisms are not imperfect versions of ideal forms; they are members of populations, each more or less adapted to its environment. It is precisely this variation that enables evolution, making it a more Aristotelian than Platonic process.

From S.J. Gould, I carry with me his extraordinary narrative ability to communicate evolutionary science through meticulously crafted essays, enriched by encyclopedic historiographic documentation. Although he was a staunch critic of the notion of progress in biological evolution—and of progress more broadly—the criteria he developed to assess its presence or absence have been invaluable to my own studies on the evolution of biological complexity.

From R. Lewontin, I have been deeply influenced by his reflections on the laws—many still undiscovered—that govern the transformation of genotypes into phenotypes and vice versa, all within the context of evolving populations. Although he was skeptical about the empirical demonstrability of many of these laws, he laid the conceptual groundwork for genomic research in evolutionary biology.

I was particularly struck by E.O. Wilson's revival of the concept of consilience, initially coined by philosopher William Whewell. It refers to the deliberate effort to create a unified framework of understanding by integrating knowledge across disciplines. Wilson embraced this idea, advocating for the integration of science, the humanities, and the arts. In his view, all three aim to provide meaning by identifying an ordered world governed by a limited set of natural laws.

I take this opportunity to strongly recommend approaching science in the same manner as these great thinkers did: always from a non-specialized perspective, drawing on sources beyond one's immediate field to generate new insights and theories.

5.13 Other Intelligences

Abraham Loeb heads the Department of Astronomy at Harvard University and, in collaboration with Shmuel Bialy, published a provocative paper in *The Astrophysical Journal Letters* (Bialy and Loeb 2018). The study proposed the possibility that an interstellar object—named "Oumuamua"—which passed rapidly through our solar system, could be a probe created by an extraterrestrial civilization. Without delving into technical details, the authors noted that both the dimensions and shape of the object—a cigar-shaped structure approximately eight hundred meters long and eighty meters wide—and its unusual acceleration when passing near energy sources such as stellar winds, suggest it could be propelled by solar sails, enabling it to travel vast distances. Since this discussion remains within the realm of science, alternative, less spectacular explanations have emerged. Ultimately, a general consensus has formed around the need to have a pre-deployed probe capable of approaching such objects upon detection—Oumuamua was detected in October 2017—and studying them up close.

I believe three considerations regarding this discovery and its interpretations are timely. The first is the need to continue expanding space research programs and to invest in advanced technologies not only for detecting extraterrestrial life—hence the earlier comment about having a probe ready to intercept objects like Oumuamua—but also for developing interstellar travel prototypes. In this regard, I strongly agree with one of Stephen Hawking's key research priorities.

The second consideration concerns extraterrestrial intelligence itself. Surprisingly, the scientific attitude toward this topic has shifted in recent years. While popular imagination has long entertained the idea of intelligent beings elsewhere in the universe, many respected scientists—Loeb among them—now argue that, based on sheer probability and the vastness of the cosmos, planets with conditions similar to Earth must exist, allowing life, including intelligent life, to evolve. More skeptical voices contend that even if intelligent life exists elsewhere, the likelihood of contact is extremely low due to the immense spatial and temporal scales involved. Nevertheless, there is a growing tendency within science to reject the notion that we are the sole intelligent beings in the universe.

The third consideration pertains to the evolution of our understanding of intelligence on Earth. We have gradually moved from viewing ourselves as the only intelligent species to recognizing that intelligence itself is an evolutionary trait. Evolutionary experiments are remarkable in this regard: organisms as different from us as particular species of octopuses exhibit clear signs of intelligent behavior (Godfrey-Smith 2016), sometimes surpassing that of many mammals.

This leads to a paradigm shift in the secular anthropocentrism that has long shaped our self-perception and our view of other entities in the universe. Now, guided by science itself, we are more inclined to accept the possibility of multiple forms of intelligence, both terrestrial and extraterrestrial. The anthropic principle suggests that the cosmological constants allow us to exist and ponder these questions. But perhaps this principle extends—or so it seems—to other intelligent entities as well.

5.14 About Accidentality

Just before the final epilogue, Stephen Jay Gould ends his book *Full House: The Spread of Excellence from Plato to Darwin* (1997) with the following blunt statement: "We are monumental accidents of an unpredictable process lacking momentum toward complexity, not the intended result of a series of evolutionary principles anxious to engender an organism that could understand the reason and mode of its necessary construction."

Gould was determined throughout his life to dismantle, on scientific grounds, the deeply rooted idea that we are the final product of a directed evolutionary process. He does not criticize the classic pre-evolutionary thesis in which humans appear as a genuine and important product of Creation, though not necessarily the highest, as other celestial entities may be superior. Instead, Gould's critique targets the notion of progressive evolution, arguing that it lacks direction and that no fundamental force has guided life from its simplest beginnings to the emergence of complex organisms. Natural selection, he insists, cannot be the driving force behind such a trajectory, as it operates over short evolutionary distances and can both increase and decrease complexity.

Gould's central argument is that it is a prejudice to view evolution as a ladder ascending from simplicity to complexity. His insightful critique urges us not to focus on how the average of any trait associated with complexity changes over evolutionary time, and even less on how these traits manifest in newly evolved organisms. Instead, he advocates for observing "the full house"—not the averages, but the variation. His most eloquent example contrasts bacteria with humans. Bacteria have evolved since the origin of life and today represent the largest biomass on Earth, thriving in virtually every imaginable environment. If we consider abundance and evolutionary success, bacteria clearly emerge as the dominant species. Our species, by comparison, is a minor appendage on the vast, bushy tree of life. "The fashion is bacterial," Gould declares. We should abandon the ladder metaphor—fish, amphibians, reptiles, mammals—as it reflects a misguided view of life's true, branching, and undirected nature.

Gould also introduces the concept of the "left wall" of evolution: the inevitable emergence of complexity from simplicity. While complexity does arise, it is not directed and remains marginal compared to the bacterial world, which has thrived by simply continuing to exist and populate the planet.

The question of progress and the evolution of complexity remains unresolved—as is often the case in science—despite the efforts of our most admired thinkers. Gould's bacterial world may be responding to its environment in ways that record information, contributing to the increasing complexity of the organisms that have emerged since. This process does not follow the much-maligned evolutionary ladder; rather, the bushy tree of life continues to grow in complexity across many of its branches. How do we measure this complexity? We do not yet know for sure. But it is worth considering an alternative reflection: that an inevitable necessity in the evolution of life may temper Gould's emphasis on accidentality.

5.15 Hybrids

The Massachusetts Museum of Contemporary Art, commonly referred to as MASS MoCA (https://massmoca.org), is situated in the city of North Adams, located in the heart of the state. I do not doubt the wisdom of bringing a project focused on the artistic avant-garde of the United States to this city. Visitors cannot remain unaffected by everything happening there.

Among the many exhibitions—some permanent, others temporary—I will focus here on one that immediately captivated me. It is the work of American artist Trenton Doyle Hancock, whose central theme is the creation of a world based on the concept of the "mound." What is a mound? Before describing it, it's important to note that Hancock constructs an existential universe parallel to the human one—he calls it the "Moundverse," which could be translated as "monticular universe" or "montiuniverse." This universe is populated by a set of entities that are, to some extent, descendants or derivatives of the mound, and among which biological, psychological, and sociological relationships of all kinds are established. It is a world that invites both rational understanding and aesthetic immersion, which, in my view, is what draws in any visitor: to enter it is to step into something different, yet not entirely distinct from the human existential universe.

It became clear to me—likely influenced by my professional background—why I was immediately drawn to Hancock's work: the dissection of the mound's internal structure, its functioning, what it feeds on, how it grows and reproduces. I quickly realized I was encountering the concept of an organic being unlike any known life form on our planet. For Hancock, the mound is a complex, plant-human hybrid.

While contemplating Hancock's explanatory illustrations of the mound, I was struck by how recurrent this type of imaginative proposal is throughout history. One need only examine the vast array of organic entities found in the mythologies of different civilizations to see that human history is filled with beings unseen in natural history. I won't delve into the religious, sociological, or anthropological reasons behind the human creation of such unnatural beings, but their existence is undeniable, and they are many and varied.

I also cannot explain why, among the many mythological hybrids I could have recalled, Hancock's work brought to mind the image of the monk-fish hybrid described by Pierre Belon, a sixteenth-century historian and naturalist. I suspect this connection arose because I reference Belon's hybrid in my lectures on the theory of biological evolution, particularly when discussing how the pre-Darwinian world was populated by imaginary beings, rather than those observed in nature—both extant and extinct. You would be amazed by the diverse and strange biodiversity that has populated our planet in the past, with forms as fantastical as those created by human imagination. What's even more striking is our preference for hybrids, combining human traits with those of other beings. Ancient Egypt and classical Greece were rich with human-animal hybrids—birds, felines, horses, to name a few. And in the same tradition, across time, we find Belon's hybrids and now Hancock's.

Our fascination with hybrids has endured. It's not just a matter of imagining them—we've also tried to create them in reality, crossing species with or without much scientific rationale. A significant portion of the flora and fauna that coexist with us today are natural hybrids. Whether in reality or imagination, we must not forget that hybridization has always been an integral part of our human experience.

5.16 "Lifeforce": Dethroned and Recovered

Vitalism has long been a discredited theory, especially in the scientific realm, though it has retained some philosophical interest. Here, I will focus on the concept of vitalism in the context of science. It cannot be accepted because it does not align with what we observe in living beings when we study them. The theory posits that life cannot be explained solely by physical or chemical forces, introducing an insurmountable divide between organic and inorganic nature. Yet, when we examine living organisms, we find nothing that suggests they operate outside the laws of physics and chemistry. This is the consensus presented in general biology textbooks.

Mainstream historiography tells us that vitalism was superseded once it became clear that there is nothing inherently special in living beings—no phenomena, reactions, or processes in the cell, the fundamental unit of life, that defy physical or chemical explanation. Renowned scientists such as Claude Bernard and Hans Driesch once argued that living matter exhibits unique behaviors that require a distinct force. Bernard referred to this as the "*vital force*", and Driesch as "entelechy." Notably, neither scientist attributed life to supernatural causes. They were deeply engaged in advancing the biological science of their time and believed that understanding living organisms required "something more."

Science occasionally entertains advanced hypotheses that later fall out of favor due to a lack of empirical support. Bernard and Driesch proposed that living beings, as material entities, required a special force absent in non-organic matter. But no evidence was ever found to support this claim. As cellular biology advanced, no such force was observed. The chemical synthesis of urea, an organic compound, was a pivotal development that undermined vitalist explanations of life. Vitalism thus became a relic theory.

Yet, it was also Pasteur who demonstrated that life arises only from life. If, as noted above, life is governed by physicochemical processes, why can't we create it from basic components? This may seem contradictory, but it is not. We lack sufficient knowledge. Science is never complete; it is a system in constant transformation, constantly revisiting its foundations and theories. The vitalists perceived something in living beings that remains valid, even if we cannot identify anything that violates physical or chemical laws. So where does the problem lie? We are still in the process of understanding how life originated—and we still lack the science to comprehend it fully.

The theoretical biologist Stuart Kauffman (2019) addresses this in his thought-provoking book *A World Beyond Physics*. This is not a metaphysical treatise. Like the other great scientists mentioned earlier, Kauffman seeks a scientific explanation for life using contemporary concepts. He introduces ideas developed over the past fifty years. Life, he argues, is a far-from-equilibrium system that emerges thanks to specific boundary conditions, new constraints, and several self-reinforcing closures that characterize behavior as a unified whole. Kauffman identifies three fundamental closures that enabled life's origin. When the right conditions were met on Earth, life emerged—and from there, voilà, it began to unfold.

Life is not a machine; it is not an automobile. An automobile cannot do what life does: build itself. This story is far from over. Vitalism is dead—but only just. We still need a comprehensive explanation of life: one that accounts for its holistic behavior without invoking forces beyond physics and chemistry. And that is precisely what science continues to pursue.

5.17 Straddling the Line Between the Necessary and the Accidental

Two scientific publications appeared in the third week of January 2020 that led me, once again, to reflect on the question of what is necessary and what is accidental in the evolution of life. On the one hand, a study published in *Nature* reported the discovery of an archaeal species living on the seabed, more than 2500 meters below the surface, that exists at the interface between prokaryotes and eukaryotes. This organism provides compelling evidence for how the first eukaryotic cell may have formed. It is the closest archaeon yet identified to eukaryotes in evolutionary terms. Some archaea similar to this one may have entangled bacteria with their tentacle-like structures, engulfed them, and incorporated them as endosymbionts capable of surviving and reproducing within the host. This, it seems, is how the eukaryotic cell originated.

The other study, published in *Science*, supports the impactor hypothesis—as opposed to the volcanist hypothesis—regarding the cause of the last great mass extinction. It suggests that the meteorite, which was over 10 kilometers in diameter and struck the Yucatán Peninsula 66 million years ago, creating the Chicxulub crater (approximately 200 kilometers wide), was primarily responsible for the extinction event that wiped out the dinosaurs. It's worth noting that other mass extinctions have occurred in which volcanic activity played a decisive role.

One cannot help but feel perplexed and awed when contemplating what these two phenomena—and others like them—mean for life on Earth. First, they highlight life's persistence and tenacity, its ability to continue and evolve despite catastrophic, life-threatening events. With bold resilience, life restarts—a kind of neo-life. Mass extinctions have caused enormous losses in biodiversity, and while the details of what was lost or what survived may be secondary, the essential fact is that life endures. We already know that life on Earth will cease in approximately ten billion years, when the Sun exhausts its energy and undergoes a supernova explosion. But that's still far off: if life began around 3.5 billion years ago, it has only completed 25% of its journey—there's still 75% to go.

Second, what has life discovered in this first stretch? Many things. It is often said that these discoveries are accidents, just like the catastrophic events that caused mass extinctions—but with opposite effects. Yet I believe they are accidents of a different kind. In fact, I would argue that life, through its evolution, creates the conditions for increasingly complex innovations to emerge across the tree of life. With a certain anthropocentric arrogance, it is often claimed that the appearance of the eukaryotic cell marks the beginning of complex life. And indeed, from it emerged plants, fungi, and animals.

Still, much remains to be investigated about how and why—if, and also why—life has evolved as it has. Across the great tree of life, when studied in all its dimensions, we find recurrent evolutionary convergences, which seem to suggest the presence of an unavoidable necessity, governed by laws we have yet to understand. Is this the same complexity I refer to when life appears on a planet?

We have only been here a short time. We still have a long way to go in determining how much of evolution is accidental and how much is necessary. And if we consider how much time remains for life on Earth, we might also think that intelligent life still has a long journey ahead—perhaps long enough to answer these fundamental questions.

5.18 The Virus Dilemma

We are currently witnessing the epidemic spread of a coronavirus outbreak, a new strain compared to those previously identified. Recently, another outbreak of the Ebola virus has emerged, and each year we monitor whether a new strain of the influenza virus will circulate. Viruses are responsible for a wide range of infections, from the common cold to HIV/AIDS. The list goes on, as cases of viral infections—both familiar and newly emerging—continue to appear worldwide. While these viruses can be classified based on their genetic similarity to known strains, outbreaks often involve new variants. However small viral genomes may be, the number of possible mutations is virtually infinite, and occasionally, one emerges that successfully proliferates and spreads. In this light, although viruses are not considered autonomous living beings, they are arguably the most Darwinian in one of life's most fundamental pursuits: survival.

We must abandon the notion that viral outbreaks and epidemics are distant events occurring in remote or exotic locations. Viruses are ubiquitous. They exist in all kinds of species, including humans, and can, with some probability, jump between hosts. When highly infectious strains emerge in humans, we must recognize that viruses can spread globally like an invisible third passenger, especially in our highly interconnected world with modern travel capabilities.

Some viruses are indeed more dangerous than others. Based on metrics such as their ability to kill the host, we can establish a ranking of virulence. However, let's consider what I mentioned above: viruses, like all living entities, aim to maximize their survival. There is no conscious intent in this process—it is simply a matter of replication. Yet viruses require other living organisms to replicate. If they kill their hosts too quickly, they risk eliminating themselves—unless they manage to transmit to another host before doing so.

This brings us to two key factors that place viruses in a fundamental evolutionary dilemma: virulence and transmission. A virus may be highly virulent, replicating aggressively and causing severe disease or death. However, this can hinder transmission if the host dies before the virus is spread. Conversely, high transmissibility is advantageous, as it ensures the virus moves from host to host; however, it often comes at the cost of lower virulence. These two traits are somewhat antagonistic.

This reflects a general principle in biology: resources are limited, and trade-offs are inevitable. For example, an organism that invests heavily in reproduction early in life may have a shorter lifespan. In contrast, a long-lived organism may reproduce more gradually but still leave a comparable number of descendants. One cannot easily be both highly fecund and long-lived.

Viruses face a similar trade-off. High virulence boosts replication within a host but may reduce transmission. High transmissibility ensures spread but may require moderating virulence. The ideal virus, from the virus's perspective, would be both highly virulent and highly transmissible, maximizing replication and spread. However, research has yet to confirm whether such viruses exist—or could exist.

The current coronavirus outbreak appears to have a notable feature compared to previous ones: asymptomatic carriers can transmit the virus. This is a highly effective strategy for the virus, allowing it to spread undetected. Importantly, its virulence does not appear to be significantly higher than that of previous strains. Nevertheless, asymptomatic transmission presents a monumental public health challenge.

References

Bialy S, Loeb A (2018) Could solar radiation pressure explain 'Oumuamua's peculiar acceleration? Astrophys J Lett 868:L1

Bostrom N (2014) Superintelligence: paths, dangers, strategies. Oxford University Press, Oxford

Dennett D (1995) Darwin's dangerous idea. Evolution and the meaning of life. The Penguin Press, London

Godfrey-Smith P (2016). Other minds: the octopus, the sea, and the deep origins of consciousness. Farrar, Straus and Giroux, New York

Goméz JM, Verdú M, González-Megías A et al (2016) The phylogenetic roots of human lethal violence. Nature 538:233–239

Gould SJ (1997) Full house: the spread of excellence from Plato to Darwin. Harmony Books, New York

Harari YN (2016) Homo Deus: a brief history of tomorrow. Harvill Secker, London

Kauffman S (2019) A world beyond physics. The emergence of evolution and life. Oxford University Press, Oxford

Kurzweil R (2006) The singularity is near: when humans transcend biology. Penguin Books, New York

Mayr E (2004) What makes biology unique?: Considerations on the autonomy of a scientific discipline. Cambridge University Press, New York

Moya A (2009) Naturaleza y futuro del hombre. Síntesis, Madrid

Nagel T (2012) Mind and cosmos: why the materialist neo-darwinian conception of nature is almost certainly false. Oxford University Press, Oxford

Penrose R (1989) The emperor's new mind. Oxford University Press, Oxford

Pinker S (2018) Enlightenment now: the case for reason, science, humanism, and progress. Viking, New York

Rovelli C (2018) The order of time. Penguin Random House, New York
Young RL et al (2019) Conserved transcriptomic profiles underpin monogamy across vertebrates. Proc Natl Acad Sci USA 116:1331–1336
Zimmer C (2001) Parasite rex: inside the bizarre world of nature's most dangerous creatures. Free Press, New York

Chapter 6
Artificial Evolution

Abstract In the last chapter, I reflect on the future of our species through the lens of artificial evolution—our growing capacity to intervene in nature and ourselves. I explore transhumanism and the philosophical, ethical, and ontological implications of modifying human biology and creating entities that emulate us. Drawing on literature, science fiction, and contemporary technological developments, I examine how artificial intelligence, synthetic biology, and algorithmic thinking challenge our understanding of consciousness, identity, and evolution. I question whether machines can ever achieve consciousness and whether we can decode the biochemical algorithms that define intelligence and self-awareness. These reflections are grounded in a broader concern: that our species may be transcended by its own creations. Through this inquiry, I aim to provoke thoughtful dialogue about the limits of science, the nature of humanity, and the responsibilities we bear as creators of a new evolutionary path. The scenarios explored in this chapter should be read as plausible futures for reflection, not as predictions or technological imperatives.

Keywords Algorithmic thinking · Artificial intelligence · Consciousness · Limits of science · Self-awareness · Systems biology · Transhumanism

6.1 Transhumanism: Artificiality and Intervention

In 2009, well before the emergence of interest in transhumanism, I published a book, *Naturaleza y Futuro del Hombre* (Moya 2009). In it, I referred to a concern that I believed was not widely shared—at least in the Ibero-American sphere: the future of our species and its eventual replacement by another entity, a singular product of our creativity. Later, I realized—after a brief review of the history of Western thought—that this concern is more widespread than I had imagined. If we delve into certain theologies, into the philosophy of classical thinkers regarding utopias—Plato, for example—into the utopians of the Renaissance, the more recent techno-utopias envisioning technological worlds or totalitarian political orders—such as Aldous Huxley—into much of the literature on science and technology, into a large

A. Moya, *Reflections on Science and Evolution*,
https://doi.org/10.1007/978-3-032-25682-9_6

part of science fiction—as in the case of Isaac Asimov—or into some futuristic films—for example, Ridley Scott's *Blade Runner*—we can quickly conclude that the idea of our species being transcended has long been present in the intelligentsia of every era. Today's technological society—cyber-connected and globalized—exhibits more transhuman and near-future elements than those of the past, which it abandons with surprising speed, as if the future were settling into the present while the past fades away.

However, reflecting on our transhumanizing actions is pertinent because it cannot be said to be a recent practice. It is something intrinsic to our species. Since our origins, we have intervened in our surroundings—what I call "the natural-other"—and in ourselves. We are a compulsively intervening species. The rational understanding or explanation of what is intervened has indeed grown continuously, particularly since the emergence of science in the West. Without realizing it, we have moved from the "artificialization" of the natural-other and ourselves—driven solely by the success of the resulting product—to rational and scientific "intervention." Such are the modern intervention technologies applied to living beings, particularly humans, such as genetic engineering or synthetic biology. And of this type, too, are the increasingly sophisticated, autonomous, and intelligent robotic products, or the technological implants that "cyborgize" our bodies.

I believe we can accept that the processes of artificialization and intervention—both in the natural-other and in ourselves—have driven the creation of new societies and civilizations. A recent example is "cyberculture," which is fostering a global society constantly nourished by new communication technologies—some of which, yet to emerge, we cannot even imagine. In our cultural evolution, artificialization and intervention have been constants, leading to technological revolutions in agriculture, industry, information, and communication—each of which has nurtured the genesis of new societies. Against this background, what makes us think we cannot transcend ourselves?

6.2 In defense of Post-humanist Thought

The history of Western thought is full of defenses of humanism—a philosophy, or rather several philosophies, whose origins can be traced back to the Italian Renaissance of the fifteenth century and which, in general terms, seeks the development of the human being's qualities as a whole. The recourse to humanism has always been a persistent appeal in the face of the barbarism and brutality with which some human beings have treated others during various tumultuous moments in history. As always happens in thought, this first humanism arises in opposition to another prevailing current of the time: scholasticism, the first regulated university education that served to train doctors, lawyers, and professional theologians, where they were taught—through textbooks approved by the Church—logic, medicine, natural philosophy, law, or theology, all of which are ultimately based on the logic and philosophy of Aristotle.

Post-humanism, today, is a current of thought that seems to conflict with humanism. Why? The subjects from which it draws are not those that have traditionally nourished the teaching of humanism, but rather stem from science and technology themselves. To a large extent, post-humanism emerges as a philosophy associated with the exponential growth of science, characterized by ever more visible and transcendent innovations that increasingly shake our lives and world at shorter and shorter intervals. The question one can ask about such a trend is the following: What will that future world be like, one that experiences an even more dizzying rate of innovation than we do now? What can we say about a world that is being intervened upon, where our nature will also be altered? And, finally, why should we shy away from reflecting on the possibility of creating thinking machine entities or even changing our ontological status in a kind of enhancement whose nature we have yet to glimpse?

The survival instinct, fed by intelligence and the sense of self, is so deeply rooted in us that any reflection on the creation of new entities or possible changes in our ontological status is considered a direct threat to humanity and, ultimately, to the values that have been explicitly championed by humanism for centuries. For example, we are reluctant to think that we might have enough willpower to renounce ourselves in favor of other entities that, being the product of our creation, might surpass us in many or all capacities. Or will we be so narcissistic as to assume that we are unbeatable? If nature and the evolution of living beings reveal anything, it is that they are frankly capable of improvement; no one has designed them, and they resort to what they have at hand to survive—in what is known, following François Jacob, as the "bricolage" of evolution. Can we envision a future in coexistence with, or even supplanted by, post-human beings—our children in reality, products of a rational and planned bricolage? Would we be willing to make such a sacrifice? Although we may abhor this approach, I believe it is necessary to consider it seriously.

The question for reflection and debate, assuming the exponential growth of science and technology, is whether we will reach the point of creating post-human entities—and whether what we want is a future human society with all the social and coexistence improvements we can imagine, or the contrary: a post-human one, about whose structure and social organization little can be said at present. In reality, there is a clear imbalance in favor of humanism because we have not yet given sufficient thought to the nature of post-human society. Hence, I advocate thinking about post-human entities. The history of humanity, despite its outstanding social and coexistence achievements, carries enormous burdens. Our species does not seem capable of subverting the confrontation and barbarism that self-destruction through war implies. Something Cainite defines us—something that, although it appears to have been instrumental in bringing us here and allowing us to enjoy the splendor of culture and civilization, has also led us toward destruction and extermination. Is it possible to think of a world that is finally not Cainite? Would it be more human?

6.3 The Tree of Life, the New Biology, and Post-evolution

It has always been maintained that Darwin's outstanding contribution to science was the discovery of natural selection as the driving force of evolution. Although its significance is unquestionable, another contribution—also the work of the British genius—is no less important: the Tree of Life. In his magnum opus, *The Origin of Species*, Darwin presents a figure illustrating the relationships between species, showing that the most recent ones have older ancestors, and those, in turn, even older ones, tracing back to the most ancestral of all, presumed to represent the common origin of all life. The shape of this figure is that of a tree—the great icon of life—and echoes the double helix structure of DNA.

Modern biology provides evidence that evolution can occur through forces beyond natural selection. Nevertheless, all scientific evidence strongly supports the tree-like nature of evolution. The philosopher of biology Elliott Sober notes that although Darwin was primarily focused on finding a causal and material explanation for the transformation of species—natural selection being his landmark discovery—his concept of the Tree of Life holds equal or even greater significance.

Modern evolutionary biology has advanced considerably. It has been invigorated, refined, and reinterpreted by emerging life sciences, including genomics, evolutionary developmental genetics, and complex systems biology. Let us consider how.

The sequencing of the genomes of a growing number of species—now numbering in the thousands—as well as of multiple individuals within representative species (including thousands of human genomes and those of model organisms such as bacteria, fungi, plants, and animals), has allowed us to consolidate and deepen our understanding of the Tree of Life. This has helped identify the forces that shape its structure and explain why it appears as it does.

Moreover, the mechanisms and intricate details of the genetic foundations responsible for the morphology of major organismal groups have been clarified, revealing which genetic changes are necessary depending on the gene type and thereby explaining how evolution has transitioned from one organismal form to another.

Finally, we have entered an era of holistic, integrative research on organisms. A longstanding biological tradition has been anti-reductionist, arguing that life cannot be understood by breaking it down into parts. Yet today, the possibility of detailed computational modeling of specific cells—and predicting their behavior—is becoming routine, precisely because reductionist biology has enabled us to identify and combine those components. This emerging field is known as systems biology.

These recent trends in biological research justify the claim that we now possess new tools to investigate evolutionary biology from a radically new perspective. Traditionally, evolutionary biology has focused on reconstructing the history of life. But can we not also begin to explore where life is headed?

While reconstructing the past remains valid, we are taking giant steps toward investigating life beyond evolution—a kind of post-evolutionary biology. Our species can reconstruct its history and trace its roots within the Tree of Life. But it can also alter the natural course of evolution by fabricating entities that have never existed before. Post-evolutionary biology is the synthetic biology of new biological entities.

6.4 Reflection on Synthetic Life

In reality—or to a large extent—when we speak of synthetic life, we are referring, in the first instance, to a declaration of intent, not to say a desire, wish, or hope, that a component of the biological entity, or the entity as a whole, will be controlled; that is, that it will function as designed and not deviate in the slightest from its assigned role. If there had not already been notable scientific advances, it would be of little use for the future survival of this biological science of new entities to continue relying solely on declarations of intent. When certain sciences advance, specific incipient achievements propel them toward new knowledge, recognition, and general acceptance by both the scientific community and society.

The fundamental aspiration is that the new object of production should function in the same way as, for example, an automobile functions after coming off the assembly line. The comparison with the automobile is as valid as with any other man-made product. The union of its components in a specific sequence along the assembly line results in an entity that functions in a certain way, because all elements are well understood and the manufacturer assembles them in a particular order, ensuring the whole works as intended. This analogy, that a manufactured biological entity should function like an automobile, leads us to some reflections I consider pertinent.

The automobile, as is well known, has an average lifespan. Any of its parts—especially those fundamental to its purpose—inevitably wear out as materials decompose, are transformed by friction, or for any of the many reasons that lead the car to cease functioning. This is not due to an accident, but simply because it wears out and no longer serves its intended purpose. To what extent does this metaphor apply to the manufactured biological entity? In fact, regarding the biological equivalent of the automobile, we can ask: Will it suffer wear and tear that leads to it stopping or functioning differently? The average lifespan of a car can be extended as long as its components are systematically maintained and repaired. This process could be extended indefinitely unless various considerations—such as economic factors—lead us to conclude that buying a new car would be preferable.

In contrast, the biological entity must consider that the capacity to extend its average life comes from within. It must possess self-intervention mechanisms. The automobile, at least with current technology, does not allow for this type of intervention. But from the synthetic biological object, we do expect it. Even if automobiles are obsolescent and can be repaired externally, biological entities address their obsolescence with varying success through self-intervention. If a synthetic biological entity is, by definition, a biological entity, it will cope with its obsolescence autonomously—or more autonomously—than an automobile can. It would only cease to do so if we were to introduce, emphasizing the synthetic aspect of the concept, some form of control that prevents such a natural dynamic from occurring. But I can conceive of no better form of control of a biological entity than one based on a deep understanding of its nature.

In other words, the intervention in a biological entity—or, failing that, the fabrication of a biological entity from biological components, i.e., the creation of a synthetic biological entity—has a fundamental difference from the entity we call an automobile: the latter is designed according to fully known principles.

On the contrary, the synthetic biological entity begins with components whose rules of construction and functioning are not entirely understood by us. And to the extent that they are not, they cast us into the realm of unpredictability. I do not mean by this that we are thrown into absolute unpredictability or that there is no form of control over what we handle. There are mechanisms of control and external interventions that can make the entity controllable—although I should rather say increasingly controllable. The synthetic biological entity is embedded within biological entities or systems that, as evolving organisms, must be sufficiently understood to control their self-interventions—science, always more science.

6.5 Human, All too Human

Although the title coincides with that of one of Nietzsche's books, it serves as a fitting introduction to the problem that some seek to highlight: the inability of our species to survive under the conditions imposed by its biology, particularly those related to social behavior. That is the thesis. I must say it is surprising coming from Luisgé Martín, who, in his recent work *El mundo feliz: una apología de la vida falsa* (2018), concludes that we have no solution unless we transcend ourselves, that we must go beyond what we are. Because we are "too human," we are cyclically confronted with limitations and structural defects that make universal coexistence difficult, if not impossible. Between Hobbes, for whom "man is a wolf to man," and Rousseau, for whom "nature has made man happy and good, but society depraves him and makes him miserable," we should think about where we stand and what we are, based on the facts that our history reveals about the difficulties of living together without thinking—consciously or unconsciously—that one comes before all the others.

In light of the evidence that science reports on human nature, we are not equidistant between Hobbes and Rousseau; in fact, we are closer to the former than to the latter. This becomes evident when we examine the kind of governance we have been giving ourselves. Through more or less implicit and assumed pacts and rules of coexistence, we attempt not to go against others, driven by the imperative of our tendency to do so.

We have always thought, in opposition to our natural humanity, that the only way to control it would be to apply some universally accepted Kant-like rule: "Do not do to anyone what you would not want to be done to you." But why such a rule? Well, because we are implicitly accepting that it is within our natural humanity to act otherwise—that what we are born with, first and foremost, are desires, which are sometimes manifested in actions that will most likely lead to harm to others, harm that we would not want to be reciprocated in any way. It was Heidegger who commented that if societies could disappear or, more specifically, if advanced forms of social organization were to collapse overnight, what we would still have left—something that we could not eliminate—would be our biology.

Well, the panorama I have presented about humanity through the theses defended by Hobbes, Rousseau, Kant, Nietzsche, and Heidegger can change in the light of science. Our humanity can be transcended. We can go ontologically beyond ourselves in what seems immovable: our biology. As Luisgé Martín rightly argues, we can embrace Huxley's *Brave New World*—we can consciously deceive ourselves, and, knowing that we are doing so, we can pursue the happiness made possible by transformations controlled through bioengineering. These next-generation drugs allow us to access desired states, or virtual realities that place us in non-ordinary worlds. In this new and happy world, work will not exist, sex will cease to be reproductive, death will not be as we know it, and virtual reality will allow us to live many lives. Our humanity feels an aversion to all this, lashes out against such pretensions, and seeks to present this new world as an atrocious dystopia where everything has been transformed for the worse—because, in this new reality, there will be no worthwhile humanity, since we will no longer be under the umbrella of the categories that characterize it: freedom, independence, and emotion. Let me go one step further. I cannot say at this moment whether the post-human future will be as anti-human as implied here, given the science we have yet to discover. I can imagine novel ontological realities, post-human entities, and a corresponding universal society that would delight our most illustrious thinkers. These reflections are intended to raise ethical questions rather than to offer definitive answers or prescriptions.

6.6 The Algorithms of Intelligence and Consciousness

An article has just been published (Silver et al. 2018) in which the authors, who work for Google's artificial intelligence division, indicate that they have created a program—the usual term now is "algorithm"—of general purpose: it can play any board game and only requires, as a starting point, the rules of the corresponding game. It begins by playing randomly, but after a sophisticated learning process based on a reinforcement strategy and playing against itself, the algorithm achieves superhuman performance—in other words, it generally plays better than the best human players of the game in question. Specifically, they convincingly demonstrate that the AlphaZero algorithm—as they refer to it—defeats world champions in chess, shogi (a Japanese variant of chess), and the Chinese game of Go.

What can we say about the AlphaZero algorithm and human intelligence? In short, it surpasses us. We cannot argue that the algorithm's ability is limited to a specific game or that it merely outperforms a human in a particular context. Here, the authors have developed a general-purpose algorithm with the unique ability to learn autonomously and ultimately play better at any competitive board game. In other words, we are approaching the development of general-purpose artificial intelligence, and we cannot exclude the possibility of creating new algorithms that, this time, have a general purpose beyond games, applied to tasks related to intelligence and other human capabilities.

For historian Yuval Noah Harari, in his *Homo Deus: A Brief History of Tomorrow* (2016), we humans, too, are made of algorithms; we are, according to him, "biochemical algorithms." In any case, the difference between the algorithms developed in artificial intelligence programs and those of biological origin is that the former are known to us, since we have built them ourselves. However, biological algorithms are far from being discovered, as they are numerous and diverse. Being products of biological evolution—or, in other words, having been shaped by evolution over billions of years—their transcription into the machine language of a computational algorithm is not immediate. We have not written them ourselves, and our work is cut out for us, even assuming we are ultimately biochemical algorithms. I do not mean that it's impossible, but it is undoubtedly complex. It occurs to me that the genome harbors algorithms encoded in its structure's various layers and folds; however, many of them have yet to be discovered. Worse still, perhaps, will be the challenge of finding the biochemical algorithms embedded in the multiple layers and folds of the brain—if we continue with Harari's suggestive metaphor.

A final thought: What is the biochemical algorithm corresponding to consciousness, assuming it gives rise to it? We have already seen that human intelligence—whose algorithm we do not yet know—may well be surpassed by machine intelligence, whose algorithms we are actively developing, as mentioned above. But will we ever be able to construct the algorithm of consciousness, as we are doing with intelligence? We cannot claim that the intelligent algorithm developed to play more efficiently than we do is conscious of its performance. It is not. It simply plays better than we do; we have achieved that. We will probably make further progress in that field, but will we be able to develop an algorithm that achieves the status of consciousness? I believe consciousness is a complex property of the mind, whose manifestation requires a multitude of layers and folds in the brain's organization—many of which remain unknown to us. It is unclear to me whether a singular moment will emerge in humanity's future when artificial intelligence research achieves a conscious algorithm, and even less so that such an algorithm would be the fortunate product of combining previously developed intelligent algorithms. Although the world that awaits us may change radically from the current one with the development of new algorithms, achieving the algorithm of consciousness remains critical to determining whether we will be replaced.

6.7 On Post-human Ethics

Much of the current reflection on our present and future relationship with digital machines seems to center on the development of norms for their control by humans. In her article *"Ética digital"* (*El País*, Friday, December 7, 2018), philosopher Adela Cortina briefly presents the fundamental norms that should regulate the relationship between humans and machines. She outlines the following five principles:

1. The principle of beneficence, whereby digitalization serves human beings and the planet.
2. The principle of non-maleficence, which involves avoiding possible harm to humans resulting from the misuse of data and, above all, the potential submission to machines.
3. The principle of individual autonomy, which can be strengthened by artificial intelligence, nonetheless requires that humans remain the ultimate decision-makers in significant matters.
4. The principle of justice, which ensures an equitable distribution of the benefits of digitization.
5. The principle of explainability, which holds that those affected by the digital world must understand it—especially when they are called upon to take responsibility.

It may seem a truism to say so. Still, these principles are approached from a humanistic perspective, fundamentally oriented toward safeguarding humanity's interests and warning us about the possible destructive effects of digital machines should they play a more significant role in society. The destruction I refer to is tied to the principle of individual autonomy that underlies humanism. It seems to provoke panic when we imagine going from being the controllers to being controlled, and, to a large extent, the above principles are designed to protect us from that scenario.

It is worth considering, however, a question that Adela Cortina also raises in her article: machines are automatons and are not responsible for their actions. For this reason, she argues, it is necessary to propose ethical codes that are both valuable and unbiased—codes aimed at safeguarding the principle of human autonomy in relation to machines. But artificial intelligence is advancing by leaps and bounds, producing algorithms—automatons, in Cortina's view—that increasingly improve upon human characteristics, starting with intelligence. It is no longer true that these algorithms are exclusively designed for specific purposes; more and more general-purpose algorithms are being developed that emulate and substantially enhance human qualities.

There is much caution—perhaps excessive caution—surrounding the principles of digital ethics. Provided that digital entities are not conscious (a subject that deserves separate reflection given its extraordinary importance), what is certain is that if machines are not mindful and yet perform most tasks better than we do, it may be worth considering the possibility of placing ourselves in their hands. Let us reflect and reconsider the principles of digital ethics in light of this thesis:

1. Beneficence: Nothing suggests this would change, since machines would not act for their own benefit but in favor of humans and the planet.
2. Non-maleficence: It is difficult to imagine harmful effects from data misuse or submission to machines if the corresponding algorithms are designed to serve us, with far better guarantees of non-maleficence, given their efficiency, than if we relied on human decisions, which are prone to error. However, this implies relinquishing our autonomy.
3. Autonomy of individuals: This is a critical issue, as noted above. While artificial intelligence may be oriented toward strengthening our most important decisions, it is also possible that this principle would no longer apply in a post-human society where humanistic individuality has disappeared.
4. Justice: In the hands of the digital world and super-intelligent machines, we might expect improved distribution of resources—far better than what has occurred throughout human history.
5. Explainability: It would be difficult to explain to all humans the vast capabilities of super-machines; we would need to trust them, as they would be intrinsically programmed to do nothing other than provide us with happiness.

I am aware of the vertigo that arises from reconsidering the principles of digital ethics in favor of a post-human digital ethics. Still, the evolution of science and technology propels us into a world where it is pertinent to reflect on whether humanism has reached its ceiling.

6.8 Schubert's "Finished" Symphony

Different media outlets now report that Huawei scientists and engineers have developed an algorithm that completes the famous and marvelous *Symphony No. 8* by Franz Schubert, better known as *The Unfinished*. The brilliant composer died at thirty-one amid a creative maelstrom. Considering his country of origin, Austria, I am convinced that today he would have had the opportunity to continue his innovative work thanks to support from public or private entities. This leads me to reflect that individual creation—fundamental to the sustainability and development of humanity—has not been well supported historically; in fact, it remains insufficiently supported. The case of Schubert calls out for the implementation of measures to ensure that no talent is lost, as has happened to many throughout history and continues to happen today. However, although we cannot and should not allow this—a double imperative—I am convinced that we are making progress in capturing and adequately supporting talent.

But let us return to the matter at hand, and let it be clear that Schubert's work does not need any finishing touch; it is what it is—the product of a comprehensive set of factors. Now, the reaction of some intellectuals and music professionals to this attempt to complete an extraordinary piece of music seems to be quite unanimous: among other things, they claim it is artificial, lacking soul, and that Schubert's creative act involved many unpredictable elements—unexpected leaps, psychological states, ethical and aesthetic considerations, even social ones—that cannot be replicated by an algorithm. They even, and not without reason, state that the algorithm "is indifferent to them." What need is there to finish a piece like this—or any other unfinished work—if that were the case?

The perspective I want to bring here is different. As with the issue of consciousness, I do not believe we are close to achieving—nor am I convinced it will be feasible—a conscious machine. But there are already algorithms that are more intelligent than we are. I have already mentioned one with general-purpose intelligence that, using neural networks and starting only from the rules of various board games, can play better than the best human experts. Why could not a piece of music, the product of an algorithm, be just as excellent and provoke the same emotions in listeners as any piece written by a brilliant human? It can be argued that a musical composition is more than the composer's intelligence, and that creativity is a human capacity closer to consciousness than intelligence. I disagree. While it is true that knowledge of particular circumstances—historical, ideological, sociological, artistic currents, etc.—helps us better understand and appreciate the greatness of what we are listening to, the truth is that the piece itself, regardless of these circumstances, produces emotions in the listener that sustain its particular greatness.

It can be argued that this is a deception—that a decontextualized piece, the product of an unknown source, lacks the same quality and cannot evoke the same emotions as a piece created by an author within a specific context that is undeniably present in the work. However, the algorithm also has its context. It does not emerge from nothing; it has a history within the field of artificial intelligence. Indeed, there have been previous attempts—both with this Schubert symphony and with many others—to orchestrate increasingly efficient algorithmic procedures so that a professional musician might conclude that the piece possesses the characteristics of a masterpiece. Why do we react so negatively to this kind of achievement? What fear does it provoke in us?

6.9 On the Future of *Homo sapiens*

It was with great expectation that I went to one of the large auditoriums at MIT (Massachusetts Institute of Technology) during my stay in Cambridge to attend a series of conferences titled "*The Future of Homo Sapiens.*" That was the name of the event. I attended a session featuring five speakers invited by Professor Joseph Paradiso, who works at the university and hosted and moderated a final roundtable discussion among them. The speakers were five well-known scientists in their respective fields: Pattie Maes in cognitive science, Max Tegmark in physics, David Sinclair in biotechnology, George Church in genetics, and Edward Boyden in neuroscience. All of them work either at MIT or Harvard University. Except for Tegmark, who delved into the future of the species and the planet, the presentations of the others were overly specialized and had uneven relevance to the topic that was supposed to unite them. In summary, the speakers discussed human enhancement, new technologies in artificial intelligence, and the use and improvement of the capabilities of other living beings.

As the event unfolded, I realized, with growing frustration, that something was missing. I became concerned when I briefly considered that the future of our species could not be left exclusively in the hands of scientists like those invited to participate in this endeavor. Then I calmed down, thinking that the event had been organized to discuss new technologies for human enhancement—and that the title did not correspond to what was actually presented. It would have been straightforward, had the organizers adopted a broader vision of the topic, to invite intellectuals from other fields to participate. I missed three issues that I consider to be of utmost relevance.

The first was the assumption that we are here to stay. None of the speakers framed their presentations around the topic they had been invited to reflect on: the future of the human species. At no point was the possibility raised that our species might disappear—for any reason—nor was there any discussion of the nature of the entities that might replace us. There was an implicit assumption in all of them, but nothing explicit, that, as I say, we are here to continue.

The second issue I overlooked was the complete absence of any general reflection on our current species and the challenges it faces—and will continue to face—in the near future, including social inequality, inter-nation inequality, and the sustainability of the planet. Inviting a philosopher, sociologist, or other humanities scholar would have been highly relevant to discuss whether humanity can have a future if these issues are not adequately addressed. I immediately thought that science, in its specialized forms, may not be the most effective tool for helping us envision the future of our species. While I believe science is essential to leading us into the future, I also recognized the need to reflect on questions that were not addressed at the event.

The third issue I missed was the political dimension. I want to reiterate that the topic was so broad and so significant that it clearly required the presence of an expert on how changes are likely to unfold in human society as a whole—and, more specifically, what kinds of social forces or activism should emerge, and in what types of collectives, to promote the changes that will allow us to continue as a species.

So, I was left hesitant at the end of the event. I thought that what had been presented with great fanfare as *The Future of Homo Sapiens* should have been titled something like *Natural Sciences and Human Enhancement*. Several essential components were missing to drive that cart forward properly.

6.10 Live Longer, but Live Well

Globally, life expectancy (the number of years a person is expected to live) and healthy life expectancy (the number of years a person is likely to live without disease or disability) increased significantly between 2000 and 2016. Life expectancy rose from 66.5 to 72 years, and healthy life expectancy from 58.5 to 63.3 years. Considering the growth of the world's population during this period, the increase in both parameters is notable. Note also the remarkable difference between the two measures: while life expectancy increased by 5.5 years, healthy life expectancy rose by 4.8 years. And the fact remains that—even if only concerning health, which is no small matter—living longer is not the same as living well.

This issue must be considered when we say that the average human lifespan is increasing with the development of civilization. It is also necessary to examine how that other measure—healthy life expectancy—is evolving. However, we must be cautious, as I am discussing human civilization as a whole, and the figures I've cited are averages across a profoundly unequal world. The poorest and least developed countries have significantly lower life expectancies than those of more prosperous and well-developed nations. We still have a long way to go before these disparities are eliminated. Let us focus first on the essential goal: to live longer, yes—but above all, to live well.

This final sentence serves as a call to attention for geroscience, a new branch of science that brings together, on one hand, biomedical researchers seeking a basic understanding of the molecular and cellular mechanisms behind aging, and on the other, geriatricians aiming to improve the quality of life for older adults. Geroscience seeks to elucidate the intricate relationship between aging and the chronic diseases commonly associated with old age. Since aging is the major risk factor for most chronic diseases, the "geroscience hypothesis" has been proposed: that the basic mechanisms of aging play a key role in the increased susceptibility of older individuals to multiple chronic conditions. The leading advocate of this new discipline and proponent of the hypothesis is Dr. Felipe Sierra, former Director of the Division of Aging Biology at the National Institute on Aging, part of the U.S. National Institutes of Health.

Understanding these mechanisms could be decisive in curing, limiting, or at least postponing the onset and severity of multiple chronic diseases. And if we succeed in understanding them, we will have the option to intervene early enough to ultimately achieve a longer healthy life expectancy—a life free from chronic diseases and the disabilities they bring. Living longer and living well could be combined into a single expression: well-being. Some might argue that longevity is irrelevant if the life lived is of high quality. But under "well-being," we could include everyone—those indifferent to longevity and those who value it. In any case, as I say: good living.

For Ortega y Gasset, this was a defining feature of our species compared to animal life. It is not merely a matter of living—that is, surviving—but of living well throughout our lives. And well-being, it must be said, is more than just living healthily, though that is no small thing. Anyone approaching or already in old age knows how important good health is to face the final years of life.

6.11 Orthodox Transhumanism

It is important to situate the question of transhumanism within broader coordinates than those that frame it as a relatively recent current of thought, emerging under the influence of scientific advances in robotics, artificial intelligence, and synthetic biology—fields that have largely, though not exclusively, developed in the United States.

Anya Bernstein, a professor in the Department of Anthropology at Harvard University, published a 2019 study on transhumanism in contemporary Russia (*The Future of Immortality: Remaking Life and Death in Contemporary Russia*). It is essential to highlight the diverse and fascinating transhumanist ideology in Russia, which can be traced back to the nineteenth century, with Nikolai Fyodorovich Fyodorov (1827–1903) as its precursor. Fyodorov was an Orthodox Christian philosopher who advocated for life extension, immortality, and the resurrection of the dead through scientific means. He is therefore considered a foundational figure in transhumanist thought.

The question of precursors is intriguing: who qualifies as a precursor? Is it someone who, at some point, proposed or demonstrated something that went unnoticed for a long time until rediscovered or revived by others years later? If that is the definition, then it is inaccurate to call Fyodorov a precursor, because his thought took root and has had followers to this day, though not in the Western West, but in Russia.

It is all too common in today's West—more precisely, in the Anglo-Saxon West—to label as "precursors" those who said or demonstrated something that seems to have only recently been discovered. The argument posits that since the idea is "new," its earlier formulation must be merely an antecedent. But that is not the case with Fyodorov. His philosophy laid the foundation for an intellectual and social movement known as cosmism, which continues to this day and is supported by the posthumous publication of his work in 1906 under the evocative title *Philosophy of the Common Cause*. After a brief search, I must note that there are very few studies in the West on Fyodorov, at least in Spanish or English, and I have not found a translation of his work in either language, although it may exist in other languages.

Fyodorov's philosophical ideas form a system in which the cosmos, Earth, and humanity are deeply interrelated. As a philosophical system, Russian cosmism can be considered an intellectual current that seeks to explain life and humanity's mission to transform and perfect it. The theses of Jesuit theologian and scientist Pierre Teilhard de Chardin (1881–1955)—widely criticized, though he has fervent followers—come immediately to mind, both within the Church and in science, particularly in Anglo-Saxon contexts. Could Teilhard de Chardin also be considered a precursor of transhumanism?

Another Russian reference in transhumanist thought—though there are others — dates back to the first quarter of the twentieth century, with Alexander Bogdanov (1873–1928), a theorist, economist, and politician, among many other roles, who developed and applied ideas of human enhancement. Bogdanov not only theorized about the rejuvenating effects of blood transfusions but also practiced them repeatedly; in fact, he is believed to have died as a result of his eleventh transfusion. He also speculated on the benefits of transfusions from older individuals to younger ones, suggesting that a mature immune system could be transferred to younger individuals. Remarkably, he developed a utopian philosophy based on the supposed benefits of this communion of blood.

It is worth noting that not only the cosmist theses developed in nineteenth-century Russia but also the socialist ideology of the now-defunct Soviet Union have underpinned and inspired the development of social movements—still active today—that advocate for transhumanist egalitarian policies from governments.

In contrast to the "Silicon Valley" transhumanism rooted in capitalist ideology, it is worthwhile to acknowledge these earlier movements, which emerged long before the American ones and are grounded in very different ideological foundations and aspirations. This should not be overlooked.

References

Harari YN (2016) Homo Deus: a brief history of tomorrow. Harvill Secker, London

Martín L (2018) El mundo feliz. Una apología de la vida falsa. Anagrama, Barcelona

Moya A (2009) Naturaleza y futuro del hombre. Síntesis, Madrid

Silver D et al (2018) A general reinforcement learning algorithm that masters chess, shogi, and go through self-play. Science 362:1140–1144

Open Conclusion

As I reach the end of this journey through science and its many entanglements—with society, philosophy, politics, and evolution—I am left with more questions than answers. That, perhaps, is the most honest tribute to science: not its certainties, but its capacity to illuminate uncertainty. I have tried to think carefully and attentively, to reflect not only on what science reveals but on how it reveals us to ourselves. In doing so, I have often felt the quiet melancholy of the scientist—aware of the vastness of what remains unknown, and of the fragility of our efforts to understand.

Science, for all its rigor and promise, does not shield us from existential doubt. It may offer explanations, but it cannot always provide meaning. And yet, I believe it is precisely in this tension—between explanation and meaning—that science becomes most human. It is not a fortress of truth, but a path we walk, often hesitantly, sometimes blindly, always with the hope that the next step will bring us closer to understanding. That hope, even when unfulfilled, is a form of grace.

Throughout these reflections, I have returned again and again to the idea that science must remain open—open to other forms of knowledge, open to critique, open to the unpredictable. It must resist the temptation of dogma and embrace the humility of approximation. In this openness lies its strength, and perhaps its salvation. For if science is to serve humanity, it must first recognize its place within it—not above, not outside, but alongside our fears, our dreams, and our contradictions.

I do not know what science will become in the years ahead. I do not know what we will become. But I trust that as long as we continue to reflect, to question, and to imagine, we will remain faithful to the spirit of inquiry that defines us. And if, in the end, we find ourselves still searching, still uncertain, then perhaps we will have done something right.

A. Moya, *Reflections on Science and Evolution*,
https://doi.org/10.1007/978-3-032-25682-9

Index

A. Moya, *Reflections on Science and Evolution*,
https://doi.org/10.1007/978-3-032-25682-9

L

M

N

GPSR Compliance
The European Union's (EU) General Product Safety Regulation (GPSR) is a set of rules that requires consumer products to be safe and our obligations to ensure this.

If you have any concerns about our products, you can contact us on

ProductSafety@springernature.com

In case Publisher is established outside the EU, the EU authorized representative is:

Springer Nature Customer Service Center GmbH
Europaplatz 3
69115 Heidelberg, Germany

www.ingramcontent.com/pod-product-compliance
Lightning Source LLC
Chambersburg PA
CBHW071245230926
51849CB00001B/237
9783032256812